Minerals, Rocks and Inorganic Materials

Monograph Series of Theoretical and Experimental Studies

Vol. 1

Edited by

W. von Engelhardt, Tübingen • T. Hahn, Aachen

R. Roy, University Park, Pa.

J. W. Winchester, Ann Arbor, Mich. • P. J. Wyllie, Chicago, Ill.

Subseries:
Experimental Mineralogy

W. G. Ernst

Amphiboles

Crystal Chemistry
Phase Relations and Occurrence

With 59 figures

SPRINGER-VERLAG NEW YORK INC. 1968

W. Gary Ernst, Ph.D., Associate Professor,
Department of Geology and Institute of
Geophysics and Planetary Physics,
University of California, Los Angeles, California, USA

ISBN-13: 978-3-642-46140-8   e-ISBN-13: 978-3-642-46138-5
DIO: 10.1007/978-3-642-46138-5

© 1968 by Springer-Verlag New York Inc.
Softcover reprint of the hirdcover 1st edition 1968
Library of Congress Catalog Card Number 67–31479.

The use of general descriptive names, trade names, trade marks, etc. in this publication, even if the former are not especially identified, is not to be taken as a
sign that such names, as understood by the Trade Marks and Merchandise Marks
Act, may accordingly be used freely by anyone.

Title No. 3731

FOREWORD

This series of monographs represents continuation on an international basis of the previous series MINERALOGIE UND PETROGRAPHIE IN EINZELDARSTELLUNGEN, published by Springer-Verlag. The voluminous results arising from recent progress in pure and applied research increase the need for authoritative reviews but the standard scientific journals are unable to provide the space for them. By their very nature, text-books are unable to consider specific topics in depth and recent research methods and results often receive only cursory treatment. Advanced reference volumes are usually too detailed except for experts in the field. It is often very expensive to purchase a symposium volume or an "Advances in . . . " volume for the sake of a specific review chapter surrounded by unrelated chapters. We hope that this monograph series will by-pass these problems in fulfilling the need. The purpose of the series is to publish, at reasonable prices, reviews and reports of carefully selected topics written by carefully selected authors, who are both good writers and experts in their scientific field. In general, the monographs will be concerned with the most recent research methods and results. The editors hope that the monographs will serve several functions, acting as supplements to existing text-books, guiding research workers, and providing the basis for advanced seminars.

August 1967

W. von Engelhardt, Tübingen
T. Hahn, Aachen
R. Roy, University Park, Pa.
J. W. Winchester, Ann Arbor, Mich.
P. J. Wyllie, Chicago, Ill.

PREFACE

The results of mineral syntheses and phase equilibrium studies of mineral systems form the basis for interpretive mineralogy and petrology, as well as for many geochemical and geophysical interpretations of the earth's interior. Applications also range widely into ceramics and materials science. However, even the most comprehensive and modern mineralogical text-books contain only a fraction of the available experimental information, and the published compilations of phase diagrams provide unevaluated diagrams with no significant comments. This subseries of monographs in the series MINERALS, ROCKS AND INORGANIC MATERIALS provides detailed reviews on the synthesis and stability of selected mineral groups, supplementing rather than repeating material in mineralogy texts. Each monograph reviews the available experimental data in historical perspective, as far as this is possible and appropriate for the mineral group, and critical evaluation of earlier work and conflicting results usually precedes presentation of the most acceptable results at the time of publication. Hopefully, these monographs will help a reader to evaluate experimental data before applying published results to his own problem; I consider this an important function, because there are many incorrect phase diagrams in the literature. Certainly, each monograph will contain the most reliable information up to the date of preparation, and literature search can confidently be limited to more recent publications. As editor, I hope that these monographs will be useful to many teachers, students, and research workers in mineralogy, geochemistry, petrology, ceramics and materials science.

August, 1967   PETER J. WYLLIE
                                       University of Chicago

TABLE OF CONTENTS

Chapter I.

INTRODUCTION AND ACKNOWLEDGMENTS

Among the common rock-forming minerals, perhaps no major group is as poorly understood as the amphibole group. This is because double-chain silicates are very complicated with regard to crystal structure, chemical variation and natural occurrence. Consequently, although many amphibole species have been distinguished, mostly on a chemical basis, it is difficult to relate the geologic occurrences to physical and chemical principles.

The wide chemical variation of the amphiboles reflects an elaborate arrangement of cation sites of contrasting size and differing surrounding anionic configurations. These structures can accommodate cations of ionic radii between about 0.40 and 1.40 Å. All major cations of the earth's crust and mantle fall in this range, thus accounting for the fact that amphiboles are "mineralogic wastebaskets." At first glance, such a situation seems dismaying, and to a certain extent remains so, even to amphibole specialists.

Nevertheless, at the present time a renewed interest in chain silicates has led mineralogists to undertake detailed structural refinements, as well as infra-red and Mössbauer studies, of many amphiboles and the chemically and structurally related pyroxenes. While amphiboles so far investigated have atomic arrangements fundamentally similar to that proposed for tremolite by B. E. WARREN in 1929, complications involving the staggering of adjacent double-chains and order-disorder relations among the various cation sites appear to be required by the new three dimensional refinements and other structural considerations. The structural complexities apparently are related both to the chemical compositions of the amphiboles, and to the physical conditions of crystallization.

This complex interrelationship among amphibole composition, structural state and the attendant physical conditions of crystallization is potentially of great value to mineralogists and petrologists, for such minerals should be very sensitive to differences in physical and chemical environments. Determination of amphibole parageneses ultimately could help unravel sequences of changes in igneous and metamorphic rock series, provided the species of amphiboles are identified.

Unfortunately, the identification of an amphibole is difficult. Conventional petrographic and x-ray methods yield perhaps half a dozen useful parameters, while independent chemical variables, each of which

1

influences these measurable parameters, may exceed 10 in an amphibole; the task is formidable, and virtually demands a complete chemical analysis. In addition, infra-red and Mössbauer analyses or x-ray powder and/or single crystal diffraction techniques may be required to determine the structural state and degree of cation ordering.

Correlation of amphibole species with physical conditions of formation is most conveniently performed by experimental phase equilibration. From laboratory determination of the stability relationships of amphiboles at controlled temperatures and pressures, extrapolation can be made to the conditions of formation of amphiboles in rocks, provided the principles of physical chemistry are kept in mind for the more complicated natural systems. Experimental phase relationships usually have been determined for simple end-member compositions, or for compositions representing binary solid solution series; hence extrapolation to the more complicated natural mineral assemblages involves second order uncertainties. This approach has been necessary because phase relationships cannot be fully understood in complicated systems prior to an elucidation of the simple; furthermore, as abundantly demonstrated by the work of N. L. Bowen and many others, principles illustrated in the simple laboratory systems commonly, although not invariably, carry over into the more complex.

An alternative method for relating amphibole species to physical conditions of crystallization requires the coexistence of mineralogical geothermometers and geobarometers in natural amphibole-bearing assemblages. If the physical conditions of crystallization for a large number of different amphibole-bearing rocks could be determined unambiguously, the observed variations in amphibole compositions and structural states might then be understood and subsequently employed to indicate physical conditions in other rocks lacking geothermometers or geobarometers. Unfortunately, few natural assemblages contain unequivocal evidence regarding conditions of crystallization; hence prospects for employing this method remain bleak for the foreseeable future.

The purpose of this short volume is to summarize some current knowledge regarding amphibole crystal chemistry, chemical variability, and experimentally determined phase relationships and, as far as possible, to correlate these features with natural occurrences. Hopefully, the obvious deficiencies in our present knowledge of amphiboles will stimulate further studies.

This summary represents elaboration on notes for a series of lectures delivered November 11–13, 1966, at a short course on chain silicates sponsored by the American Geological Institute. I wish to thank the other instructors of this course, D. E. Appleman, F. R. Boyd, G. M.

BROWN, G. V. GIBBS and J. V. SMITH, as well as scientists attending the lectures; the informative discussions which characterized the sessions have influenced the contents of this volume to a considerable extent.

My appreciation especially goes to G. V. GIBBS, M. C. GILBERT, H. J. GREENWOOD, E. HELLNER and B. E. LEAKE, who reviewed the manuscript. Thanks are also expressed to those workers who sent me preprints of their amphibole investigations.

Lastly I would like to express my gratitude to Mrs. OPAL KURTZ, who drafted all the illustrative material, Mrs. LENORE AAGAARD, who typed and retyped the manuscript, and to my wife, CHARLOTTE, whose patience with "amphibolism" exceeds that of any petrologist.

Chapter II.

CRYSTAL CHEMISTRY OF THE AMPHIBOLES

Historical Review

The first structural analysis of an amphibole was carried out by WARREN (1929) on tremolite. A year later, WARREN and MODELL (1930) determined the related structure of anthophyllite. Until fairly recently, these two works were the only ones available regarding the atomic arrangement of the double-chain silicates. After WHITTAKER (1949) and ZUSSMAN (1955) elucidated the structures of a crocidolite (fibrous sodic amphibole, in this case magnesioriebeckite) and an actinolite respectively, interest in this group of minerals revived among mineralogists. More recent refinements include the following:

six hornblendes	—HERITSCH *et al* (1957, 1960), HERITSCH and KAHLER (1960), HERITSCH and RIECHERT (1960);
tremolite	—ZUSSMAN (1959);
grunerite	—GHOSE and HELLNER (1959), FINGER and ZOLTAI (1967);
cummingtonite	—GHOSE (1961), FISCHER (1966);
arfvedsonite	—KAWAHARA (1963);
two synthetic amphiboles	—PREWITT (1964);
synthetic proto-amphibole	—GIBBS (1965);
riebeckite	—COLVILLE and GIBBS (1965);
glaucophane	—PAPIKE and CLARK (1967).

Although these refinements have uncovered structural details which differ slightly from relationships elucidated by WARREN, his pioneering studies have been corroborated impressively by the modern investigations. As elucidated by GIBBS (1965) however, synthetic protoamphibole is a new structural type in which the stacking sequence of double-chains contrasts with both the common orthorhombic structure exemplified by anthophyllite, and the monoclinic structure typified by tremolite.

General Considerations

The generalized structure common to all amphiboles is shown in Fig. 1. This diagram is entirely schematic, and in detail does not apply to any

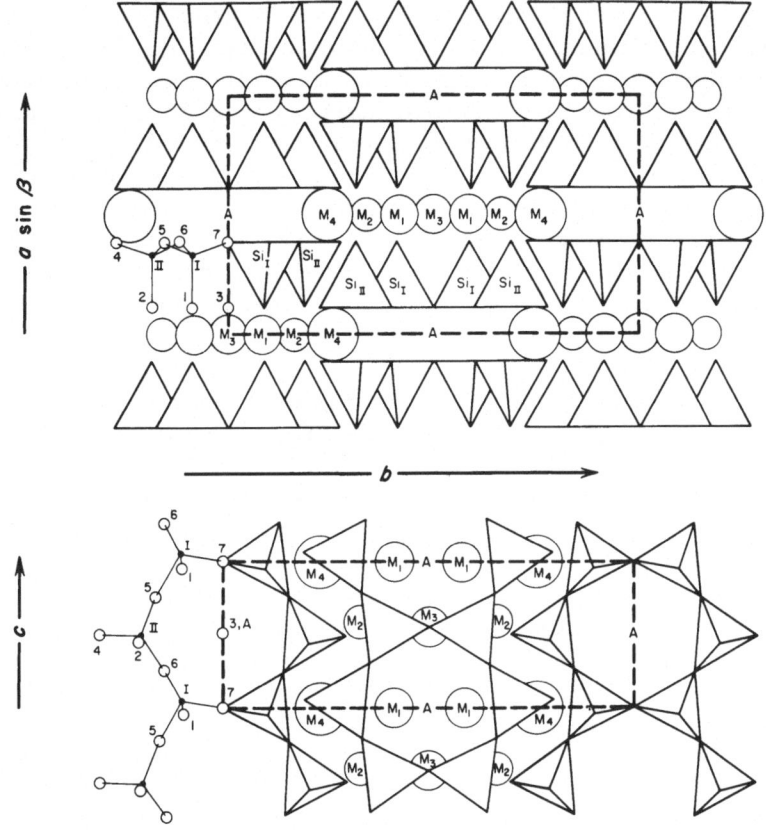

Figure 1

Schematic (001) and (100) projections of the monoclinic amphibole crystal structure modified after Colville et al. (1966). In the upper diagram, projection is normal
to the c axis. In the lower diagram, projection is normal to $a \sin \beta$; M_3 sites coincide
with the cell corners, and with the center of the cell ($I2/m$ orientation) or the centers
of the (001) cell faces ($C2/m$ orientation). Double-chains are shown in perspective ignoring the warping away from (100); backs of the chains (O_4, O_5, O_6, O_7)
are closer to the reader in the central part of the lower diagram, apical oxygens
(O_1, O_2, O_3) face the reader in the proximity of (010). The two different tetrahedral sites are designated Si_I and Si_{II}; tetrahedrally coordinated cation positions
are shown as small solid circles in the left-hand portions of the diagrams. Six-
fold coordinated cations are represented by the large circles, the Y cations in
sites M_1, M_2 and M_3, 6–8 fold coordinated X cations by even larger circles representing the M_4 site. The 10–12 fold position is indicated by the letter A. Anion
positions are shown only on the left-hand portion of the diagram by small open
circles numbered 1 through 7; the O_3 anion is generally OH or F. Orthoamphiboles
have similar structures except that (100) of monoclinic amphiboles becomes a
mirror plane; this results in a new orthorhombic cell with an $a \sin \beta$ value twice
that of the monoclinic cell.

particular species. However, the figure does illustrate the basic arrangement of constituent ions and tetrahedra in amphiboles. In the initial structural analysis of an amphibole, WARREN (1929) described the atomic arrangement in terms of a body centered cell, thus the space group was indicated as I2/m; most recent investigators have preferred the (001) face centered cell, with equivalent space group C2/m. In this paper, cell dimensions are based on the C-centered cell. In the discussion which follows, elements involved in the amphiboles are referred to as ions for convenience and consistency, but it should be remembered that the various bonds actually are intermediate between ionic and convalent types.

The amphibole structure is characterized by chains of tetrahedrally coordinated cations, crosslinked to form paired polymers of infinite length parallel to the c crystallographic axis. The basic building blocks, the tetrahedra, share alternately two and three oxygens between consecutive tetrahedrally coordinated cations in the chain. As seen from Fig. 1, each "link" of the chain consists of a six member tetrahedral ring, surrounding a central hole.

Oxygens shared by pairs of 4-fold coordinated cations are termed bridging oxygens. Backs of the chains consist of nearly coplanar oxygens; the sites conventionally are designated as the non-bridging O_4, and bridging O_5, O_6 and O_7. The latter anions are bonded to tetrahedrally coordinated cations only (unless the A, or "vacant position" soon to be described, is occupied), but O_4 is located at the periphery of the chain, and is bonded to a cation of 6–8-fold coordination. Apical oxygens lie in the next, nearly coplanar anionic layer. They occupy the non-bridging O_1 and O_2 sites, and are bonded on one side to 4-fold coordinated cations, on the other to 6–8-fold coordinated cations. Also located within this anionic layer is an anion positioned directly over the double-chain hole; anions located in this site, O_3, are bonded only to 6-fold coordinated cations (unless the A site is occupied, in which case non-polar anions such as fluorine in O_3 might conceivably be bonded to such cations—see GIBBS and PREWITT, 1966). Accordingly, this O_3 anion is typically monovalent, either OH^- or F^- except in oxidized varieties where it is O^{-2}, in contrast to the divalent oxygens which occupy all other anionic sites.

Cation structural sites consist of the following: the tetrahedrally coordinated Si_I and Si_{II}, so called because these positions were occupied exclusively by silicon in WARREN's original investigation of tremolite; octahedrally coordinated M_1, M_2 and M_3; 6–8-fold coordinated M_4; and A, an approximately 10–12-fold coordinated site which may or may not be occupied. Si_I and Si_{II} cations are located in tetrahedral interstices between the oxygen layer which constitutes the back of a double-chain and the apical oxygen layer.

The M positions are situated between two layers of apical oxygens: M_2 and M_4, located at the margins of adjacent opposite facing chains, provide the attractive forces which bind the chains together parallel to the a and b crystallographic axes. The M_2 and M_4 sites correspond to the 6–8-fold coordination positions in pyroxenes; in contrast, the M_1 and M_3 sites located well within the cation layer and flanked on both sides by apical oxygen layers are roughly equivalent to octahedral cation sites in the micas and other layer-lattice silicates.

In some amphiboles, the b dimension of the M strip of cations exceeds that of the neighboring apical oxygen layers, but for simplicity such relations are not illustrated in Fig. 1; the double-chains in these amphiboles therefore tend to be warped convexly away from this central cation strip due to structural misfit ($e.g.$, see WHITTAKER, 1949; COLVILLE and GIBBS, 1965).

A large cavity arises from near superposition of the holes of adjacent back-to-back double-chains. This A structural position corresponds approximately to the interlayer site occupied by potassium in micas, and can accommodate a large cation; however, because of the staggering of adjacent double-chains, the effective size of the site is less than that in layer-lattice silicates.

Site Occupancy

Amphiboles may be described adequately by the structural formula $W_{0-1}X_2Y_5Z_8O_{22}(OH,O,F)_2$, which represents one-half the atoms in the unit cell; W represents 10–12-fold coordinated cations occupying the A structural site, X stands for six- or eight-fold coordinated cations positioned in M_4, Y indicates 6-fold coordinated cations in M_1, M_2 and M_3, and Z refers to tetrahedrally coordinated cations in Si_I and Si_{II}. For amphiboles in which the A site is unoccupied, this vacancy will be designated by the symbol o, thus $oX_2Y_5Z_8O_{22}(OH,O,F)_2$. Occupancies of the various sites are controlled by size requirements and by the necessity of local charge balance. Names and formulae of principal amphiboles discussed in this summary are listed in Table 1.

Anions are almost invariably oxygen in the O_1, O_2, O_4, O_5, O_6 and O_7 positions. The O_3 site generally contains monovalent OH (or F) as necessitated by the condition that this anion is not bonded to Z cations of the tetrahedral chains. In amphiboles, hydrogen deficiency may arise through replacement of OH by O, with concomitant substitution of a

TABLE 1

NAMES AND FORMULAE OF AMPHIBOLE "END-MEMBERS"

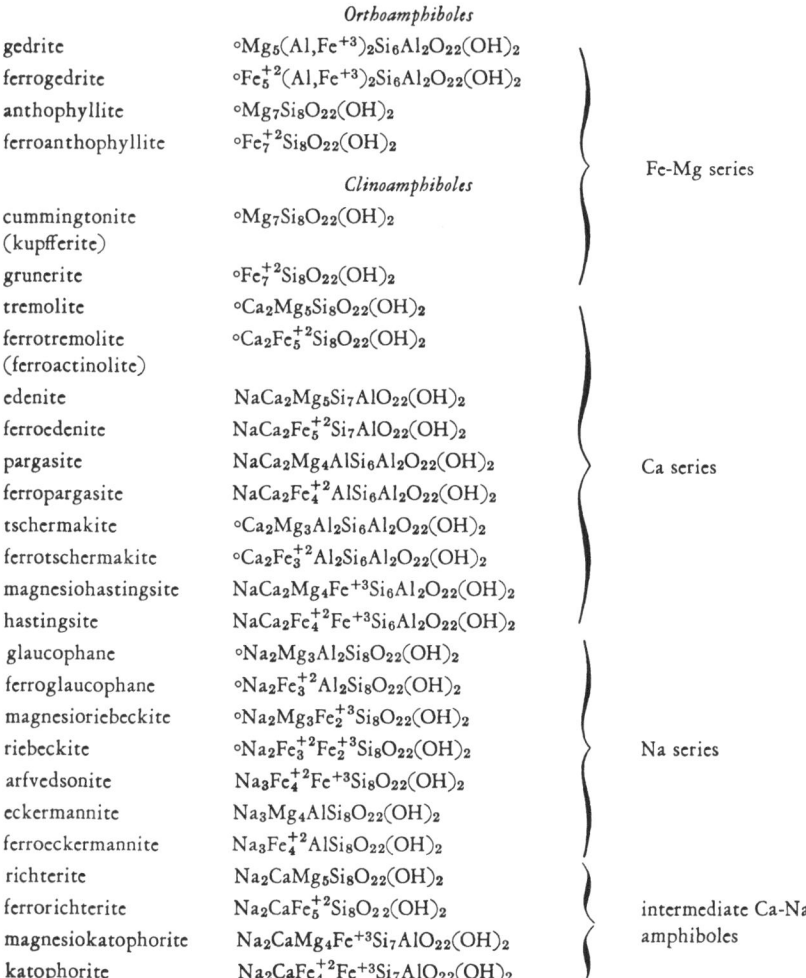

Orthoamphiboles

gedrite	$\circ Mg_5(Al,Fe^{+3})_2Si_6Al_2O_{22}(OH)_2$	
ferrogedrite	$\circ Fe_5^{+2}(Al,Fe^{+3})_2Si_6Al_2O_{22}(OH)_2$	
anthophyllite	$\circ Mg_7Si_8O_{22}(OH)_2$	
ferroanthophyllite	$\circ Fe_7^{+2}Si_8O_{22}(OH)_2$	

Clinoamphiboles   Fe-Mg series

cummingtonite (kupfferite)	$\circ Mg_7Si_8O_{22}(OH)_2$
grunerite	$\circ Fe_7^{+2}Si_8O_{22}(OH)_2$
tremolite	$\circ Ca_2Mg_5Si_8O_{22}(OH)_2$
ferrotremolite (ferroactinolite)	$\circ Ca_2Fe_5^{+2}Si_8O_{22}(OH)_2$
edenite	$NaCa_2Mg_5Si_7AlO_{22}(OH)_2$
ferroedenite	$NaCa_2Fe_5^{+2}Si_7AlO_{22}(OH)_2$
pargasite	$NaCa_2Mg_4AlSi_6Al_2O_{22}(OH)_2$
ferropargasite	$NaCa_2Fe_4^{+2}AlSi_6Al_2O_{22}(OH)_2$
tschermakite	$\circ Ca_2Mg_3Al_2Si_6Al_2O_{22}(OH)_2$
ferrotschermakite	$\circ Ca_2Fe_3^{+2}Al_2Si_6Al_2O_{22}(OH)_2$
magnesiohastingsite	$NaCa_2Mg_4Fe^{+3}Si_6Al_2O_{22}(OH)_2$
hastingsite	$NaCa_2Fe_4^{+2}Fe^{+3}Si_6Al_2O_{22}(OH)_2$

Ca series

glaucophane	$\circ Na_2Mg_3Al_2Si_8O_{22}(OH)_2$
ferroglaucophane	$\circ Na_2Fe_3^{+2}Al_2Si_8O_{22}(OH)_2$
magnesioriebeckite	$\circ Na_2Mg_3Fe_2^{+3}Si_8O_{22}(OH)_2$
riebeckite	$\circ Na_2Fe_3^{+2}Fe_2^{+3}Si_8O_{22}(OH)_2$
arfvedsonite	$Na_3Fe_4^{+2}Fe^{+3}Si_8O_{22}(OH)_2$
eckermannite	$Na_3Mg_4AlSi_8O_{22}(OH)_2$
ferroeckermannite	$Na_3Fe_4^{+2}AlSi_8O_{22}(OH)_2$

Na series

richterite	$Na_2CaMg_5Si_8O_{22}(OH)_2$
ferrorichterite	$Na_2CaFe_5^{+2}Si_8O_{22}(OH)_2$
magnesiokatophorite	$Na_2CaMg_4Fe^{+3}Si_7AlO_{22}(OH)_2$
katophorite	$Na_2CaFe_4^{+2}Fe^{+3}Si_7AlO_{22}(OH)_2$

intermediate Ca-Na amphiboles

cation by a species of higher positive charge, (*e.g.*, as in oxyhornblende—see BARNES, 1930); hydrogen excess might occur either through proton fixation by oxygens in addition to those residing in O_3, or by the presence of H_3O^+ in the A position (*e.g.*, see NICHOLLS and ZUSSMAN, 1955). Variation in proportion of the anions is extremely limited compared to cation substitution, hence compositional range in the amphiboles is usually discussed in terms of cations only.

The nature and degree of ordering among cations in double-chain silicates is currently a subject of considerable interest among mineralogists. The small size of quadrivalent silicon restricts the occurrence of this cation to sites Si_I and Si_{II} of four-fold coordination; any deficiency of Si is made up by the presence of tetrahedral aluminum (and in a few instances minor Fe^{+3}). Order-disorder among Z cations has not yet been clarified. Nor has the degree of ordering in A been established. However, modern crystal structure refinements (*e.g.*, WHITTAKER, 1949; ZUSSMAN, 1959; GHOSE and HELLNER, 1959; GHOSE, 1961; KAWAHARA, 1963; PREWITT, 1964; GIBBS, 1965; COLVILLE and GIBBS, 1965; FISCHER, 1966; PAPIKE and CLARK, 1967; FINGER and ZOLTAI, 1967) have demonstrated partial ordering among the M sites as follows: (1) M_4 always contains the highest proportion of the largest, or X cations; (2) Y cations of intermediate size are distributed nearly randomly in M_1 and M_3, although in some three dimensional refinements, a tendency for larger cations to be fractionally enriched in M_3 is discernible; and (3) smallest Y cations tend to be concentrated in M_2.

Ionic Substitutions and Lattice Effects

The perpendicular distance from Si_{II} to the *a-c* plane, the mirror plane at the margin of the unit cell, is a measure of double-chain width parallel to the *b* crystallographic axis (see Fig. 1). Typical geometries of the chains and of the Si_{II}-mirror distances for calcic, sodic and Fe-Mg amphiboles of known structure are presented in Fig. 2. Evidently the width of the chain parallel to *b* is practically independent of amphibole composition, with a minimal value of 3.05 Å for crocidolite versus a maximal value of 3.08 Å for actinolite. It does not follow, however, that the length of the unit cell along *b* remains nearly constant, merely that the chain width does not vary significantly. Obviously other structural adjustments must control the *b* repeat distance. Chain length, the *c* repeat, ranges from 5.27 Å for tremolite to 5.35 Å for grunerite. Chain thickness in the *a* sin β direction also involves only minor differences. Thus variations in dimensions of the double-chains are slight; instead they behave as relatively rigid, inflexible portions of the amphibole structure. The sizes of the large A cavities are structurally controlled by the double-chain dimensions, so these positions are geometrically invariant, too.

Accordingly, major contrasts in amphibole unit cell dimensions must reflect size differences among M site occupants. Because the mica-like structural sites, M_1 and M_3, are situated well within the double-chains, increase or decrease of the mean size of Y cations can influence only the

$a \sin \beta$, and to a minor extent the c dimension as will be mentioned later. The most strikingly variable dimension of the amphiboles is the b axis, and this length must be controlled by chain-linking X and Y cations located at the chain margins in the pyroxene-like sites M_2 and M_4 (COL-VILLE et al., 1966).

(a)

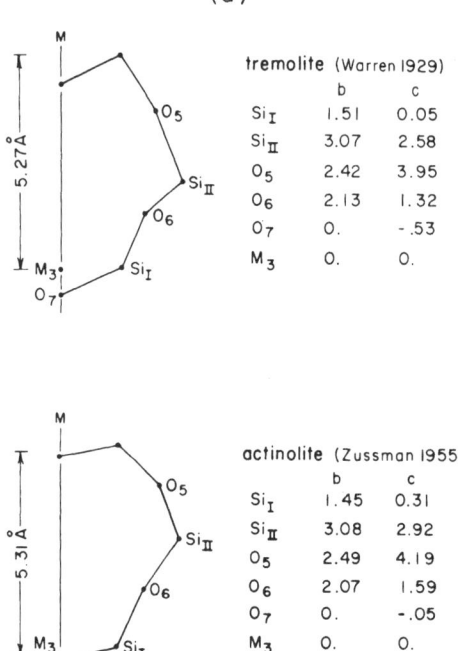

Figure 2

Projection of amphibole double chains on (100) after Colville et al. (1966). The (010) symmetry plane is normal to the plan of projection and passes through M_3 and O_7 sites, as seen from Figure 1. Coordinates are given in angstroms relative to M_3 as origin. (a) Comparison of tremolite and actinolite ($Mg_{60}Fe_{40}^{+2}$) chains; (b) comparison of magnesioriebeckite ($Mg_{86}Fe_{14}^{+2}$) and riebeckite chains; (c) comparison of cummingtonite ($Mg_{64}Fe_{36}^{+2}$) and grunerite ($Mg_{33}Fe_{67}^{+2}$) chains.

Unit Cell Dimensions of Synthetic Amphiboles

The relationship between cell parameters and chemical compositions of monoclinic amphiboles has been studied principally by COLVILLE et al (1966) who synthesized important end members and collected data from the literature; all currently available data, including the synthetic or-

(b)

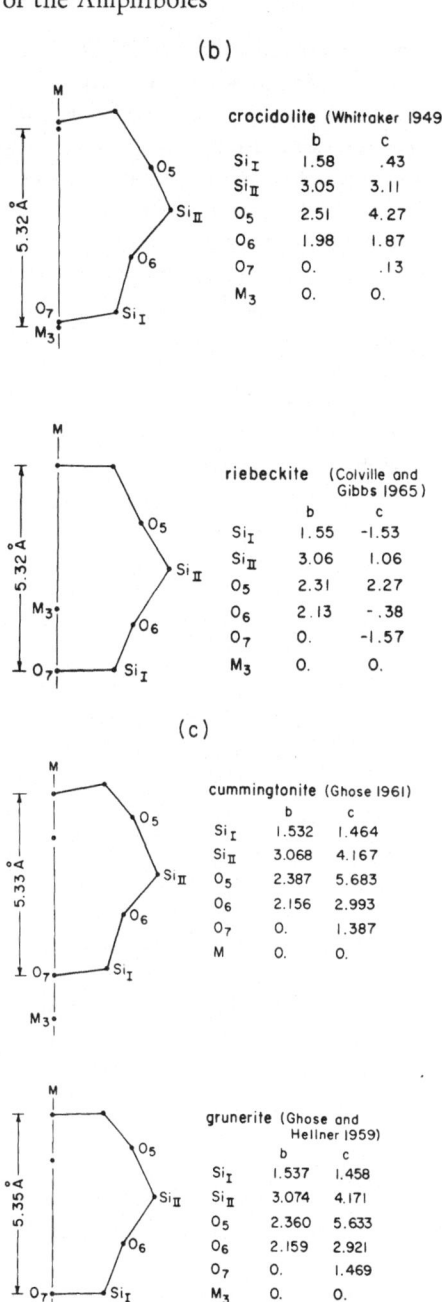

crocidolite (Whittaker 1949)

	b	c
Si_I	1.58	.43
Si_{II}	3.05	3.11
O_5	2.51	4.27
O_6	1.98	1.87
O_7	0.	.13
M_3	0.	0.

riebeckite (Colville and Gibbs 1965)

	b	c
Si_I	1.55	-1.53
Si_{II}	3.06	1.06
O_5	2.31	2.27
O_6	2.13	-.38
O_7	0.	-1.57
M_3	0.	0.

(c)

cummingtonite (Ghose 1961)

	b	c
Si_I	1.532	1.464
Si_{II}	3.068	4.167
O_5	2.387	5.683
O_6	2.156	2.993
O_7	0.	1.387
M	0.	0.

grunerite (Ghose and Hellner 1959)

	b	c
Si_I	1.537	1.458
Si_{II}	3.074	4.171
O_5	2.360	5.633
O_6	2.159	2.921
O_7	0.	1.469
M_3	0.	0.

thoamphiboles, are summarized in Table 2*. This approach for studying synthetics has the great advantage of permitting independent evaluation of the effects of many of the compositional variables which complicate relationships in natural amphiboles. Iron, magnesium, aluminum and silicon proportions were fixed experimentally. By noting the changes in properties of amphiboles which differ chemically by only one constituent, it is possible to discern a relatively unambiguous cause-effect relationship. With knowledge of the crystal structure and the presumed distribution of atoms in the different structural sites, the variations of physical properties can be correlated with site occupancies. Furthermore, the quantitative differences between pairs of end-members should prove helpful in predicting the properties of intermediate members. The four most useful results of COLVILLE *et al.* are summarized below.

(1) The b parameter can be related to the sizes of the cations in M_4 and M_2. Because the occupancy of M_4 is nearly constant in any one synthetic series, the control on b shifts to the occupants of M_2. Nevertheless, it is interesting to note that synthetic anthophyllite, in which M_4 is Mg, displays the same b-axis length as edenite, fluoredenite and fluortremolite, in which M_4 is Ca. (Of course it is true that both the coordination polyhedron about Mg in the M_4 site and the double-chain stacking sequence in anthophyllite differ from the corresponding relations in calcic amphiboles.) The b-axis variation involves an increase of up to about 0.33 Å as the M_2 occupancy shifts from all Mg to all Fe^{+2}. This is very nearly the increase predictable by the substitution of two Fe^{+2} ions for two Mg ions, the difference in diameters of these cations being 0.16 Å according to AHRENS (1952) and GREEN (1959). If M_2 is assumed to be occupied exclusively by Al, b decreases about 0.30 Å from its Mg value in synthetic amphiboles. This decrease is about half that predicted from the difference between ionic diameters, and suggests that either the assumption of perfect ordering in M_2 is not correct, or that chains are prohibited from more closely approaching one another by other lattice forces. Fe^{+3} occupancy of M_2 yields approximately the same value as Mg, which is of similar size. Obviously many combinations are possible and one may only conclude with some certainty that where $b = 18.30$ Å, M_2 must be occupied by Fe^{+2} and where $b = 17.70$ Å, M_2 must be occupied principally by Al.

(2) In general, the b axis length increases at a rate equivalent to four times the computed mean ionic radius increment in M_2, thus demon-

*Although synthetic amphiboles formed the basis for this study, natural analogues approaching end member compositions can have unit cell dimensions remarkably similar to the experimentally produced phases (*e.g.*, see the recent study by BORG, 1967, dealing with sodic amphiboles).

TABLE 2

LATTICE PARAMETERS OF SYNTHETIC AMPHIBOLES, SPACE GROUP C2/m

	a, Å	b, Å	c, Å	β, °	$a \sin \beta$	V, Å^3
°$Ca_2Mg_5Si_8O_{22}(OH)_2$, tremolite, BOYD (1959) and COLVILLE et al. (1966)	9.833 ±0.005	18.054 ±0.009	5.268 ±0.004	104.52 ±0.07	9.52	905.3 ±1.0
°$Ca_2Mg_5Si_8O_{22}F_2$, fluortremolite, COMEFORO and KOHN (1954)	9.78	18.01	5.27	104.5	9.47	899
°$Ca_2Fe_5^{+2}Si_8O_{22}(OH)_2$, ferrotremolite, ERNST (1966) ave. of 10	9.87	18.34	5.30	104.5	9.59	939
$NaCa_2Mg_4AlSi_6Al_2O_{22}(OH)_2$, pargasite, BOYD (1959) and COLVILLE et al. (1966)	9.906 ±0.010	17.986 ±0.017	5.265 ±0.008	105.30 ±0.14	9.51	904.7 ±1.9
$NaCa_2Fe_4^{+2}AlSi_6Al_2O_{22}(OH)_2$, ferropargasite, GILBERT (1966)	9.95	18.14	5.33	105.3	9,60	928
$NaCa_2Mg_4Fe^{+3}Si_6Al_2O_{22}(OH)_2$, magnesiohastingsite, COLVILLE et al. (1966)	9.925 ±0.015	17.982 ±0.030	5.289 ±0.011	105.61 ±0.12	9.56	909.1 ±2.8
$NaCa_2Fe_4^{+2}Fe^{+3}Si_6Al_2O_{22}(OH)_2$, hastingsite, COLVILLE et al. (1966)	9.979 ±0.027	18.152 ±0.063	5.325 ±0.027	105.20 ±0.34	9.58	930.8 ±5.97
$NaCa_2Mg_5Si_7AlO_{22}(OH)_2$, edenite, COLVILLE et al. (1966)	9.853 ±0.015	18.005 ±0.011	5.236 ±0.015	104.40 ±0.35	.51	899.8 ±1.0
$NaCa_2Mg_5Si_7AlO_{22}F_2$, fluoredenite, KOHN and COMEFORO (1955)	9.85	18.00	5.28	104.8	9.52	905
$NaCa_2Fe_5^{+2}Si_7AlO_{22}(OH)_2$, ferroedenite, COLVILLE et al. (1966)	9.999 ±0.010	18.217 ±0.021	5.314 ±0.014	105.50 ±0.17	9.59	932.8 ±3.0
°$Na_2Mg_3Fe_2^{+3}Si_8O_{22}(OH)_2$, magnesioriebeckite, ERNST (1963a) ave. of 7	9.73	17.95	5.30	103.3	9.47	901

TABLE 2

LATTICE PARAMETERS OF SYNTHETIC AMPHIBOLES, SPACE GROUP C2/m (continued)

	a, Å	b, Å	c, Å	β, °	$a \sin \beta$	V, Å³
°$Na_2Fe_3^{+2}Fe_2^{+3}Si_8O_{22}(OH)_2$, riebeckite, ERNST (1962) ave. of 15	9.73	18.06	5.33	103.3	9.47	913
$Na_{2.4}Fe_{4.9}^{+2}Fe_{0.7}^{+3}Si_{7.7}Fe_{0.3}^{+3}O_{22}(OH)_2$, riebeckite-arfvedsonite, ERNST (1962) ave. of 7	9.85	18.15	5.32	103.2	9.59	926
°$Na_2Mg_3Al_2Si_8O_{22}(OH)_2$, glaucophane I, ERNST (1963a) ave. of 10	9.75	17.91	5.27	102.8	9.50	897
°$Na_2Mg_3Al_2Si_8O_{22}(OH)_2$, glaucophane II, ERNST (1963a) ave. of 8	9.64	17.73	5.28	103.6	9.37	877
°$Mg_7Si_8O_{22}(OH)_2$, anthophyllite, GREENWOOD (1963)	18.61 ±0.02	18.01 ±0.06	5.24 ±0.01	90	18.61	1756
$Na_2H_2Mg_5Si_8O_{22}F_2$, Mg-fluoramphibole, PREWITT (1964)	9.650 ±0.005	17.920 ±0.002	5.270 ±0.005	102.9 ±0.2	9.406	888.3
$Na_2H_2Co_5Si_8O_{22}(OH)_2$, Co-amphibole, PREWITT (1964)	9.832 ±0.005	18.088 ±0.002	5.299 ±0.005	103.0 ±0.2	9.580	918.2
$Na_2Mg_6Si_8O_{22}F_2$, "fluormagnesiorichterite," GIBBS et al. (1962)	9.677	17.914	5.274	102.95	9.429	890.8
$Na_2CaMg_5Si_8O_{22}(OH)_2$, richterite, PHILLIPS and ROWBOTHAM (1967)	9.892 ±0.005	17.958 ±0.007	5.263 ±0.002	104.28 ±0.03	9.586	906.0 ±1.0
$Na_3Mg_4AlSi_8O_{22}(OH)_2$, eckermannite, PHILLIPS and ROWBOTHAM (1967)	9.762 ±0.005	17.892 ±0.011	5.284 ±0.006	103.17 ±0.05	9.505	898.6 ±0.8

strating the direct one-to-one correlation between ionic size and cell dimension (see also Fig. 3a). Of course, the b axis of the amphibole structure spans two double-chain widths, but it does not follow that four M_2 sites are encountered traversing the b repeat. Staggering of adjacent double-chains requires that the mean size of all M_2 cations increase so as to avoid structural warpage, if the two M_2 sites in any one M cation strip are occupied by larger ions. Said another way, any line parallel b in the structure intersects no more than two M_2 sites per unit repeat.

(3) As the mean size of the cations in the M positions increases, $a \sin \beta$ increases. An $a \sin \beta$ of 9.3 Å corresponds to exclusively small cations in these sites and a value of 9.6 Å corresponds to all large cations. Tetrahedral occupancy also influences $a \sin \beta$ but to a minor extent. The increase in the size of M_1 and M_3 is reflected in an increase of c, the chain length, as shown in Table 2. The increase in c as all of the magnesium is replaced by iron is about 0.05 Å. The chains are kinked for small cations such as magnesium in these sites and unkink to accommodate larger ones such as ferrous iron.

(4) In all cases where larger ions are placed in the structure, the unit cell volume increases. It is true, of course, that increased (monovalent) cation occupancy of the A or "vacant" site is only achieved by another concomitant substitution of the type R^{+n} replacing R^{+n+1}.

Strictly speaking, the conclusions stated above apply only to the synthetic amphibole end-members. Properties of various intermediate solid solutions can be very similar for many different chemical substitutions because the parameters are a function of the effective sizes of ions in specific sites in the structure. It seems probable that the effective sizes just referred to may be closely correlated with "average" cationic sizes; this last quantity of course depends on the actual degree of order attained by the amphibole in question. The nature of the X cations occupying M_4 apparently is reflected by the value of β; WHITTAKER (1960) and COLVILLE and GIBBS (in preparation) have shown that, among the monoclinic amphiboles, $\beta = 101.8-102.7°$ for amphiboles in which iron and magnesium are located in M_4, $\beta = 103.3-104.0°$ for the amphibole group in which the X cations are predominantly Na, and $\beta = 104.5-106.1°$ for the amphiboles containing Ca in M_4. Iron and aluminum contents may be indicated by the b repeat if these elements are present in large amounts but generally the relations are ambiguous. Optical properties can also give approximate iron contents. It is difficult to distinguish hornblendes of complex chemical compositions from one another, not because the effect of various substitutions are unknown but because quite different substitutions can give the same effect.

Observed *b* Axis Variation and M₂ Occupancy
of Natural Monoclinic Amphiboles

The study of synthetic amphibole end-members has demonstrated that the *b* crystallographic axis is the cell dimension most sensitive to composition. Variation of this lattice parameter as a function of the calculated mean ionic radius of cations occupying M_2 is shown in Fig. 3a, largely after COLVILLE *et al.* (1966). Sodium-bearing Co-amphibole and sodium-bearing Mg-fluoramphibole described by PREWITT (1964) have *b* axes shorter than predicted from the straight line relationship, suggested in Fig. 3a by COLVILLE *et al.* (1966). PREWITT stated that his amphiboles may depart somewhat from the stated formulae reproduced in Table 2, so perhaps the computed mean sizes of the cations in M_2 are slightly in error.

Synthetic glaucophane I and II contain two octahedral Al cations per 24 anions, synthetic pargasite and ferropargasite contain one Al^{VI} and two Al^{IV}. All other synthetic amphiboles studied are aluminum-free, excepting edenite, ferroedenite and fluoredenite, which carry a single tetrahedrally coordinated aluminum. It has been previously demonstrated by COLVILLE *et al.* (1966), that the $Al^{IV} \leftrightarrow Si$ substitution does not affect the length of the *b* axis significantly.

Computed mean radii for the M_2 cations in synthetic amphiboles are shown in Fig. 3a employing two alternative assumptions: complete order; and complete disorder. In either case, the average M_2 ionic size for a specific amphibole was obtained by summing the individual products of the radius of each ion times its mole fraction in M_2. COLVILLE *et al.* (1966, p. 1745) previously demonstrated rather unequivocally that synthetic glaucophane II is approximately 50 percent more ordered than glaucophane I, that riebeckite which crystallized at temperatures below about 500° C contains minor Fe^{+2} (on the order of 27 percent) in the M_2 site, and that riebeckite-arfvedsonite solid solution contains a much higher proportion of ferrous iron than does riebeckite. These authors were influenced by the apparent ordering observed in natural analogues and by the systematic relationship among synthetic Al^{VI}-free amphiboles, so they postulated cation ordering in M_2; however, synthetic aluminous amphiboles fail to conform to this simple picture, as they noted (COLVILLE *et al.*, 1966, pp. 1746, 1749–51). Assumption of disorder among cations in these sites results in an imperfect but more nearly straight line relationship between the *b*-axis length and the computed mean size of M_2 occu-

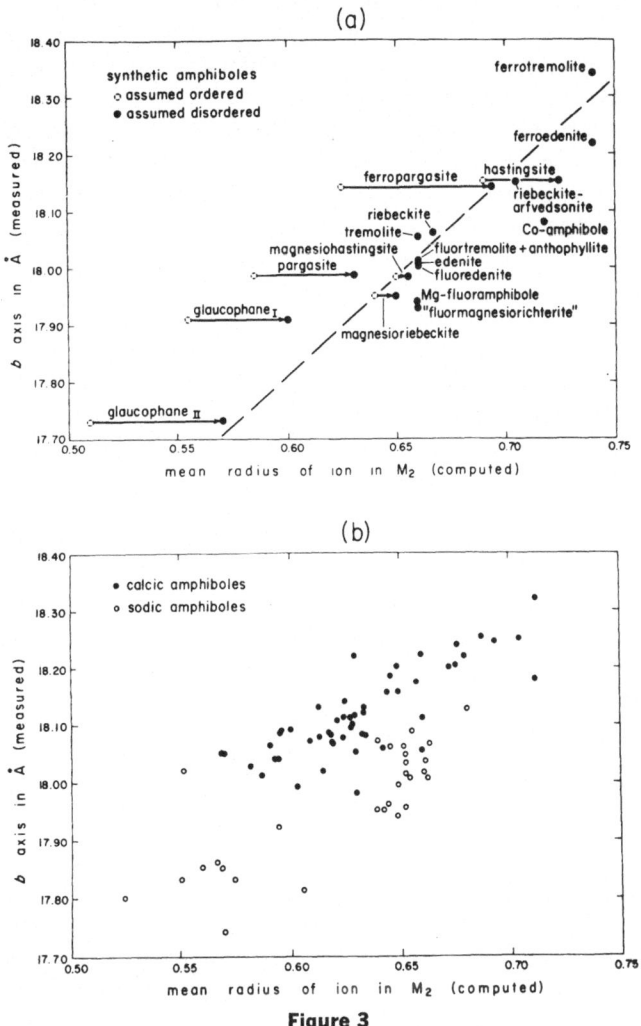

Figure 3

(a) Variation of *b* crystallographic axis for synthetic clinoamphiboles as a function of the calculated mean ionic radius of the cations occupying M_2. Glaucophane II, the hastingsites and pargasites were assumed as 100% ordered, glaucophane I 50% disordered, riebeckite 27% disordered (open circles) by Colville *et al.* (1966); the alternative and favored assumption considers M_2 occupants partly disordered in riebeckite and riebeckite-arfvedsonite solid solutions and glaucophane II, completely disordered in all other synthetics (closed circles) as discussed in text. (b) Variation of *b* crystallographic axis for natural clinoamphiboles as a function of the calculated mean ionic radius of the cations occupying M_2. The smallest six-fold coordinated cations were assumed to be ordered preferentially in this site.

pants, as seen from Fig. 3a*; this latter assumption obviously does not change the systematics previously demonstrated for end-member amphiboles lacking octahedrally coordinated aluminum and ferric iron, for in such cases, site occupancy ambiguity does not exist. Moreover, judging from experimental studies in general, the relatively high temperature, low pressure syntheses of silicates result typically in disordered phases. Hence, it is tentatively concluded that among high temperature, low pressure synthetic amphiboles, disordering of cations in M_2 is the rule; ordering is appreciable only at low temperatures (e.g., riebeckite) and/or high pressures (e.g., glaucophane II, which is approximately 50 percent ordered).

The corresponding b-axis variation of 50 chemically analyzed calcic, and 29 sodic amphiboles is shown in Fig. 3b. The compositions of amphiboles depart somewhat from the "end-member" formulae listed in Table 1; in Fig. 3b calcic and sodic amphiboles are distinguished on the basis of the dominant X cations. Mean ionic radius of the cations in M_2 was computed by summing to 2 cations per 24 anions, in sequence, octahedral Al, Ti, Fe^{+3}, Li, Mg and Fe^{+2} as necessary, assuming complete preference for this site of the smallest cations.

The correlation of b-axis repeat and M_2 occupancy among natural monoclinic amphiboles is direct, and compares with variation of synthetic end-members presented in Fig. 3a. Members of the cummingtonite-grunerite series have not been shown on the diagram. Cummingtonites exhibit considerable variation in b-axis length, even though they contain sufficient Mg to completely fill M_2; possibly these Fe-Mg clinoamphiboles exhibit differing degrees of order-disorder among X and Y cations. The distinct separation of calcic and sodic amphiboles shown in Fig. 3b depends, of course, on the calculated mean ionic radius of the cations residing in M_2. Calcic amphiboles plotted are, with four exceptions, aluminous hornblendes; in contrast, the sodic amphiboles vary principally from crossites to riebeckites and arfvedsonites, and contain only minor amounts of Al. It is probable that the behavior of aluminum in amphiboles is more complex than supposed. By analogy with the synthetic Al-rich and Al-deficient amphiboles, ordering in M_2 probably is not as pronounced as assumed. In any case, these ambiguities among multicomponent natural amphiboles must await three-dimensional crystal structure analyses for more definitive solutions.

---

*Unit cell dimensions for synthetic richterite and eckermannite (PHILLIPS and ROWBOTHAM, 1967) became available after illustrative material for this summary had gone to press, but the b axes of these two newly synthesized amphiboles support the approximately linear relationship shown in Fig. 3a.

Chapter III.

COMPOSITIONAL VARIATIONS
OF THE AMPHIBOLES

Classification

The amphiboles may be divided conveniently into three groups, based on the identity of the X cation, that is, the principal occupant of the M_4 site: Fe-Mg; Ca; and Na. Of these three groups, calcic amphiboles predominate, both with regard to the number of species recognized and absolute abundance. Sodic amphiboles are moderately widespread, while iron-magnesium amphiboles have a rather restricted occurrence.

Of prime importance is an elucidation of compositional variation within each group, and determination of the extent of mutual solid solution among the groups. The first problem is treated by graphical plots of relatively pure ternary amphiboles which may be described in terms of two independent compositional variables. The second problem is addressed by studying occupancy of M_4 and compositions of coexisting amphiboles in an attempt to delineate solvi.

The compositional diagrams which follow (principally after COLVILLE *et al.*, 1966) were prepared from analyses of Fe-Mg, calcic and sodic amphiboles collected from the literature. Chemical formulae were computed on the basis of 46 negative charges per formula ($=23$ oxygen) assuming one $H_2O(=2OH^-)$. This method of calculation tries to correct for the dubious reliability of gravimetric analyses for H_2O^+; hence the determined water is disregarded and the computation performed under the assumption of one $H_2O(=2OH)$ per formula unit (MIYASHIRO, 1957). This method may obscure real cation variation, as among the oxyamphiboles where hydrogen deficiency is compensated for by increase in valency of other cations, or by more complete occupation of the A site. However, because of variable degrees of filling for site A among amphiboles, ambiguity of structural formula will continue to exist no matter what method is applied. The only solution appears to be provided by accurate measurements of unit cell volume and density independently; consideration of these data along with a good chemical analysis unambiguously gives the absolute numbers of atoms within the unit cell, and therefore a definitive structural formula.

Iron-magnesium amphiboles to be plotted contain Ca<0.96 (all but three contain Ca < 0.31 per formula unit), Mn < 0.35 and Ti < 0.30 per formula unit. Calcic amphibole analyses utilized contain less than 0.29 ions Ti per calculated chemical formula, Mn < 0.28, (Na+K) < 1.5, (F+Cl) < 0.30 and negligible amounts of unusual components. For sodic amphiboles, extraneous ions per formula unit were: Ti < 0.25, Mn < 0.30, (Ca+K) < 0.50, Na > 1.35 and (F+Cl) < 0.30. In all cases, the number of Si ions per formula unit are within ±0.25 of the apparent graphical value. The data are presented on nine diagrams, Figs. 4–12, to show the ranges of composition among relatively *pure* members of various solid solution series. The calcic amphibole diagrams are essentially expansions of the pseudobinary side lines and the tremolite-pargasite diagonal of the plots of HALLIMOND (1943) and BOYD (1959) and various plots of SUNDIUS (1946). The purpose of the new diagrams is to show the significant natural variations between some iron- and magnesium-bearing end-members, whereas most previous investigators combined Fe^{+2} and Mg and showed compositional distributions among types varying in $(Mg+Fe^{+2})$, (Na+K) and Al contents only. For the end member formulae and terminology, the reader is referred to these compositional quadrilaterals, as well as to Table 1.

Many amphiboles could not be illustrated on diagrams such as Figs. 4–12 because they contain appreciable amounts of other constituents, or have compositions intermediate to several of the end-members considered; among such species are the richterites and the subcalcic hornblendes, in which $1.00 < Ca < 1.50$ per formula unit. The importance of these plots lies in the fact that they show relatively clearly the compositional range of large numbers of amphiboles; combination of independent chemical variables has been largely avoided. Unfortunately, because only two independent degrees of freedom can be diagrammed conveniently, compositional ambiguity results because amphiboles exhibit multicomponent chemical variability. Thus the vacant area near edenitic compositions shown by HALLIMOND (1943) is more populated in some of the new diagrams such as Fig. 5 because the Na content alone, not the sum of Na+K+Ca is plotted (Ca contents of these edenitic amphiboles are at the minimal level of acceptability, ≈1.50 per formula unit); moreover, as demonstrated by LEAKE (1962) intermediate amphiboles of the edenite-ferroedenite series do exist even employing HALLIMOND's method of representation.

The fact is that the end-member concept is of limited utility (WHITTAKER, 1966) in considering the compositional variation of the chemically complex amphiboles, because many real formulae, that is, atomic proportions, cannot be derived from the classical substitutions; moreover the procedure for calculating end-members is arbitrary and ambiguous and gives an illusion of the existence of amphibole "molecules." What is of importance is that charge balance and ionic proportions be clearly indicated in a structural formula.

Iron-Magnesium Amphiboles

Members of the cummingtonite-grunerite and anthophyllite-ferroanthophyllite series are chemically the simplest amphiboles. These are adequately represented by the formula $°(Mg,Fe^{+2})_2(Mg,Fe^{+2})_5Si_8O_{22}$-$(OH,O,F)_2$; however, orthorhombic aluminous varieties, collectively termed the gedrites, are also fairly common. As shown by Fig. 4, chemical variation among the orthorhombic amphiboles is essentially continuous from anthophyllite to both gedrite and ferrogedrite; the existence of ferroanthophyllite has not been documented. Monoclinic amphiboles range from fairly near cummingtonite (the Mg end-member has also been called kupfferite) to the $°Fe_7^{+2}Si_8O_{22}(OH)_2$ end-member, grunerite, although slightly aluminous varieties are not rare; manganous grunerites also have been described (KLEIN, 1964).

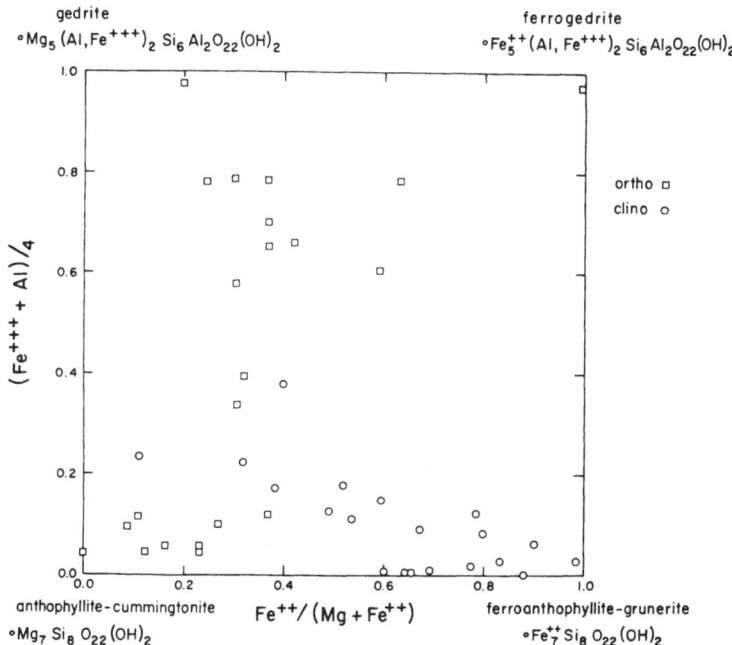

Figure 4

Distribution of natural amphiboles in the compositional range: anthophyllite (cummingtonite)–ferroanthophyllite (grunerite) gedrite–ferrogedrite.

Calcic Amphiboles

The calcic amphiboles edenite, tremolite, tschermakite and the hornblendes magnesiohastingsite, pargasite and iron-bearing equivalents are
members of complicated solid solution series. This is reflected in their

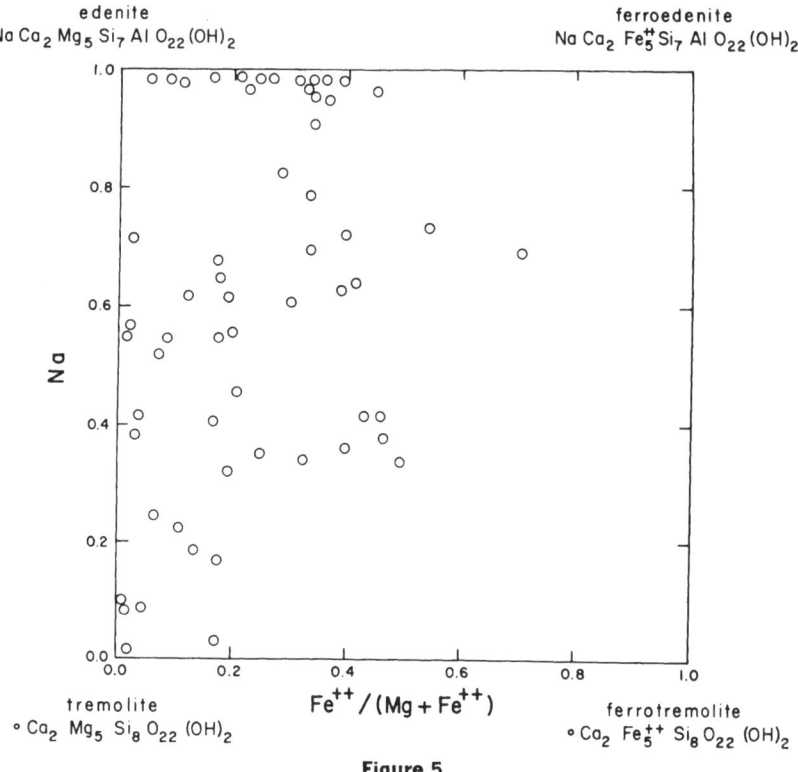

Figure 5

Distribution of natural amphiboles in the composition range: tremolite–ferrotremolite–edenite–ferroedenite (Colville *et al.*, 1966).

Figure 6

Distribution of natural amphiboles in the composition range: tremolite–ferrotremolite–pargasite–ferropargasite (Colville *et al.*, 1966).

Figure 7

Distribution of natural amphiboles in the composition range: tremolite–ferrotremolite–tschermakite–ferrotschermakite (Colville *et al.*, 1966).

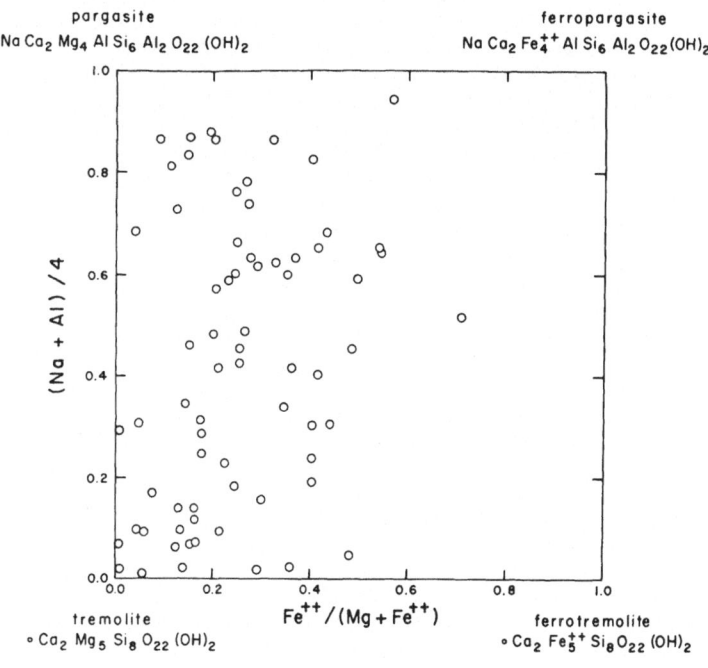

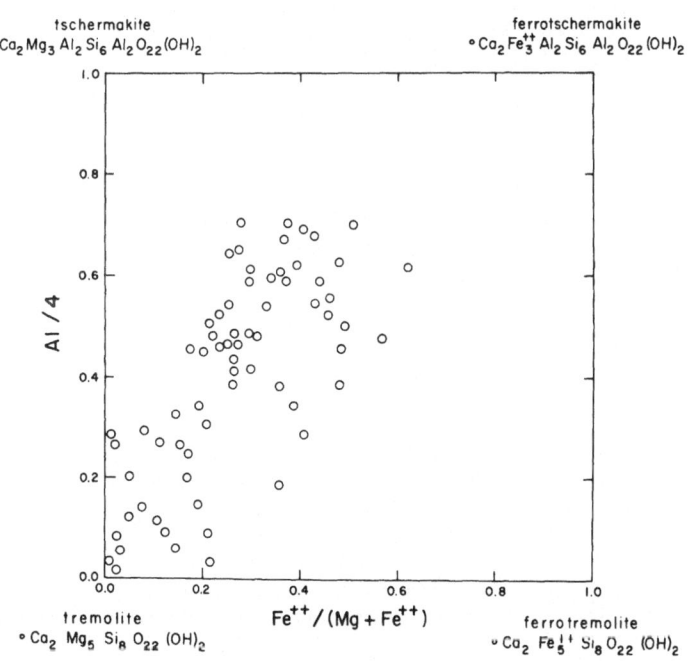

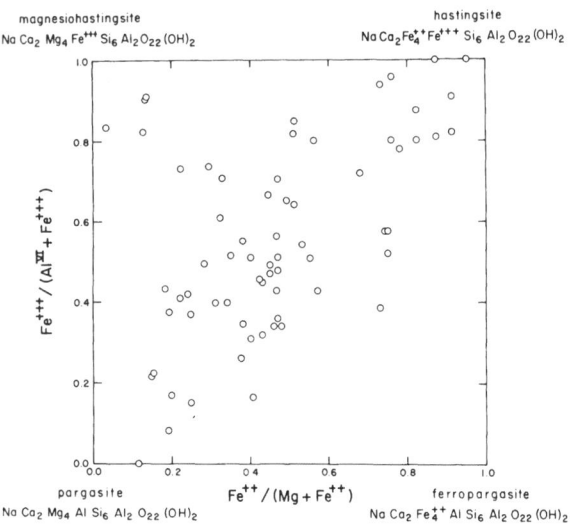

Figure 8

Distribution of natural amphiboles in the composition range: pargasite–ferro-pargasite–magnesiohastingsite–hastingsite (Colville *et al.*, 1966).

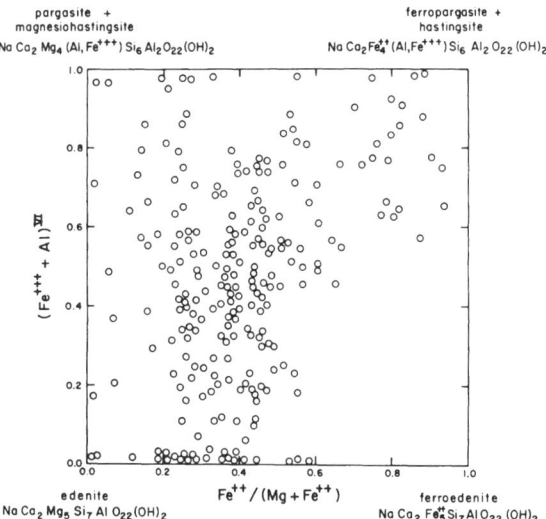

Figure 9

Distribution of natural amphiboles in the composition range: edenite–ferroedenite–(pargasite + magnesiohastingsite)–(ferropargasite + hastingsite) (Colville *et al.*, 1966).

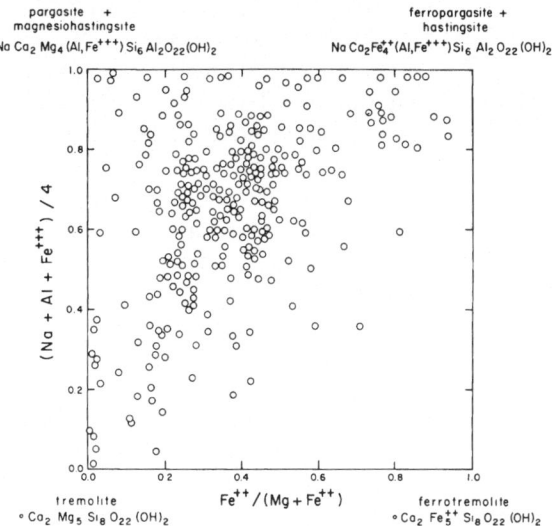

Figure 10

Distribution of natural amphiboles in the composition range: tremolite–ferro-tremolite–(pargasite + magnesiohastingsite)–(ferropargasite + hastingsite) (Colville *et al.*, 1966).

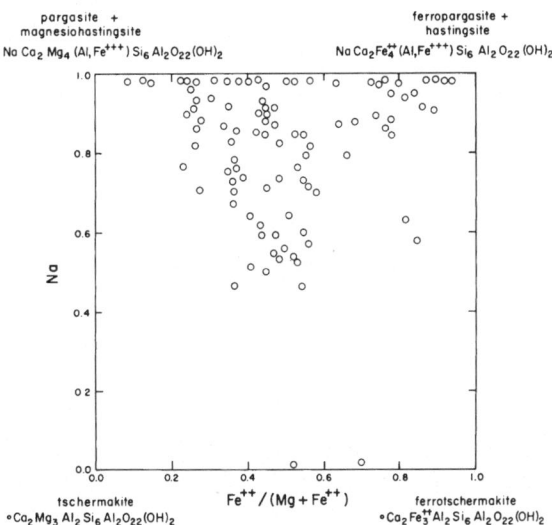

Figure 11

Distribution of natural amphiboles in the composition range: tschermakite–ferro-tschermakite–(pargasite + magnesiohastingsite)–(ferropargasite + hastingsite) (Colville *et al.*, 1966).

variable chemical compositions: $(Na,K)_{0-1}(Na,Ca,Mn^{+2},Mg,Fe^{+2})_2(Mg,$ $Fe^{+2},Fe^{+3},Mn^{+2},Ti,Al)_5(Si,Al,Fe^{+3})_8O_{22}(OH,O,F)_2$. We refer to inter-mediate members of the tremolite-ferrotremolite series as actinolites, to complex Na-Al calcic amphiboles collectively as hornblendes. Because of their chemical complexity most hornblendes cannot be expressed graphically in the familiar two or three component composition dia-grams. Not only do the number of components exceed three, but the substitution of some ions can take place in dissimilar structural sites giving rise to different properties. For example, iron can be found in tetrahedral, octahedral and perhaps eight-fold sites; the site Fe occupies as well as its valence state affects the physical properties of the amphi-bole and probably even the occupancy of other sites. Chemical variations of the calcic amphiboles are shown in Figs. 5–11.

As seen from Figs. 5–7, where ferric iron-bearing compositions are excluded, natural variations are restricted to relatively Mg-rich amphi-boles approaching the compositions of tremolite, edenite and pargasite, but not tschermakite; however, few amphiboles closely approach edenite in all compositional respects. Fig. 8 exhibits the chemical range among the sodic hornblendes. No natural hornblendes close to the purely Fe^{+2}-bearing ferropargasite have been described, but chemically similar am-phiboles containing both ferrous and ferric iron may approach $\Sigma Fe/(Mg+\Sigma Fe)$ ratios of 1.0. Figs. 9–11 demonstrate that common horn-blendes carry abundant pargasite + magnesiohastingsite, ferropargasite + hastingsite, edenite, and tremolite end-members, but only minor amounts of the ferroedenite, ferrotremolite, tschermakite and ferrotscher-makite end-members; in other words, occurrences of purely ferrous Ca-amphiboles in nature have not been recognized.

Sodic Amphiboles

Sodic amphiboles are somewhat less complicated. With the exception of arfvedsonites and katophorites which contain tetrahedral aluminum, rarely minor four-fold coordinated ferric iron and alkali metal ions in the A site, most sodic amphiboles can be referred to the glaucophane and magnesioriebeckite series with the formula $°(Na,K,Ca)_2(Mg,Fe^{+2},Fe^{+3},Al)_5Si_8O_{22}(OH,O,F)_2$, as shown in Fig. 12. Sodic amphiboles are less common than calcic analogues, but chemical variation is virtually complete among the three end-members glaucophane, magnesioriebeckite and riebeckite; compositionally intermediate members of this series are termed crossites. Amphiboles closely approaching the composition of ferroglaucophane apparently do not occur in nature.

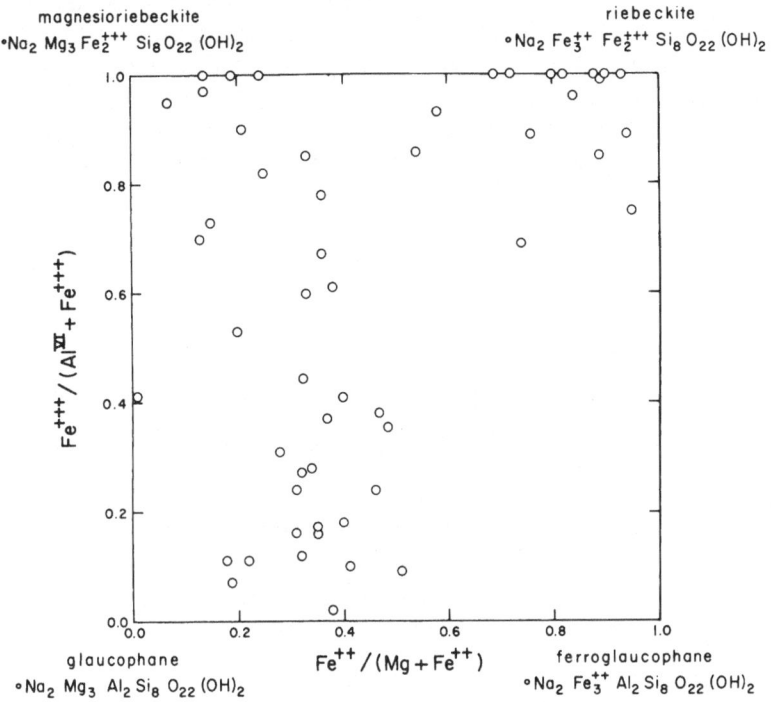

Figure 12

Distribution of natural amphiboles in the composition range: glaucophane-ferro-glaucophane-magnesioriebeckite-riebeckite (Colville *et al.,* 1966).

Amphibole Miscibility Gaps

At this point it is appropriate to mention the coexistence of certain pairs of amphiboles. Provided such associations were formed stably they indicate the existence of solvi, or prohibited amphibole compositional ranges. Three-amphibole compatibilities are unknown to the writer, but coexisting pairs include Ca-(Fe+Mg), Ca-Na and rarely, Na-(Fe+Mg) amphiboles; thus the three major groups of amphiboles are indeed chemically distinct, and separated from one another by two phase-fields. In addition, occurrences of two calcic amphiboles provide evidence of a solvus within the Ca group, and the association of cummingtonite with anthophyllite or gedrite demonstrates the presence of a solvus within the compositional range of the Fe-Mg group,—again assuming that the assemblages are stable.

Of course, the extent of solid solution between coexisting solvus phases is a function of the physical conditions of crystallization. In general, higher temperatures favor increased miscibility, perhaps even complete solid solution. Increment in pressure will increase the extent of solid solution provided there is a negative ΔV of mixing; where ΔV_{mixing} is positive, miscibility is diminished at elevated pressures. Hence, under some geologic environments, a two-amphibole assemblage may be favored for a specific rock bulk composition, while under a different P-T regime only one amphibole is produced.

Two-amphibole assemblages reported in the literature and thought to represent equilibrium associations include the following:

anthophyllite + hornblende
    (RABBITT, 1948; TILLEY, 1957);
anthophyllite + actinolite
    (RABBITT, 1948; PIRANI, 1952);
anthophyllite + cummingtonite
    (ESKOLA and KERVINEN, 1936; RABBITT, 1948);
gedrite + hornblende
    (BUGGE, 1943; SORENSEN, 1955; SUBRAMANIAM, 1956);
gedrite + cummingtonite
    (ESKOLA and KERVINEN, 1936; COLLINS, 1942; PRIDER, 1944);
cummingtonite + magnesioriebeckite
    (KLEIN, 1966);
cummingtonite + hornblende
    (ESKOLA, 1950; SEITSAARI, 1952; SHIDO, 1958; WATTERS, 1959;
    VERNON, 1962; GUNDERSON and SCHWARTZ, 1962);
cummingtonite + actinolite
    (SIMPSON, 1932; GUNDERSON and SCHWARTZ, 1962);
grunerite + actinolite
    (MUELLER, 1960; KRANCK, 1961; KLEIN, 1966);
actinolite + hornblende
    (COMPTON, 1958; SHIDO, 1958; MIYASHIRO, 1958; SHIDO and MIYA-
    SHIRO, 1959);
actinolite + crossite
    (SUZUKI, 1930; DE ROEVER, 1947; DENGO, 1950; SEKI, 1958; VAN
    DER PLAS, 1959; LEE et al., 1966; ERNST and SEKI, in press);
hornblende + riebeckite
    (HUANG, 1959).

In Fig. 13, presumed cation occupancy of the M_4 position among the three major groups of amphiboles is illustrated. This diagram was constructed assuming that all calcium resides in M_4, the deficiency, if any, being made up by Na as available, lastly by Fe^{+2} and Mg; it is misleading to the extent that actual site occupancy departs from that assumed.

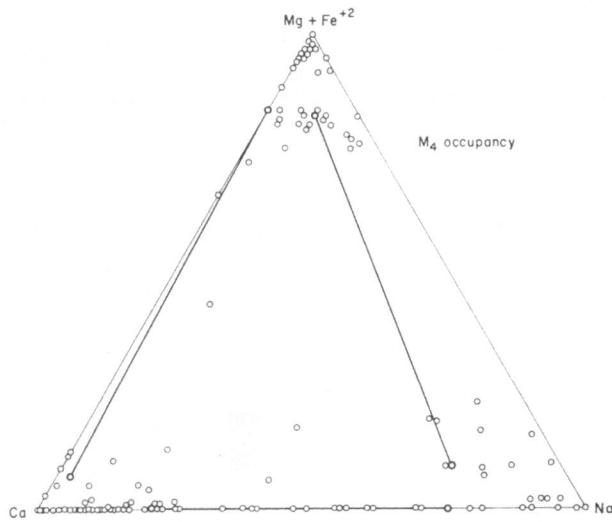

Figure 13

Presumed cation occupancy of the amphibole M_4 site. This position was computed as filled in the order Ca first, then Na and finally Fe^{+2} and Mg until a total occupancy of 2.00 per formula unit was reached. Typical pairs of coexisting amphiboles are indicated by the tie lines.

Nevertheless, three dimensional structure refinements support the presumption that large cations, and particularly Ca, are strongly fractionated into this site; however, the ordering is not perfect. The manganese contents were ignored under the assumption that this element enters all groups of amphiboles and so characterizes none.

Compositional distinction among the three major groups of amphiboles is reasonably well shown. However, the nature of the ions occupying all of the other structural positions and physical conditions of crystallization as well, undoubtedly influence the cation variation presumed within M_4; because these additional complications have not been considered, relations depicted in Fig. 13 have only semi-quantitative value at best.

Richterites, although relatively poor in alumina, are compositionally intermediate between sodic and calcic amphiboles, and so obscure the solvus relations. However, as richterites are confined to rocks which crystallized at high temperatures such as alkalic igneous bodies and scarns, they may possibly represent hypersolvus Na-Ca amphiboles.

An apparent compositional gap among the iron-magnesium amphiboles seems to be unrelated to amphibole symmetry: certain anthophyllites and gedrites are moderately calcium- or sodium-rich, although others are lacking in these constituents; similar compositional variations are found among members of the cummingtonite-grunerite series.

The only published element partitioning study for coexisting amphiboles presently available is that by MUELLER (1960, 1962). Analyses were performed by emission spectrograph, which does not allow dis-

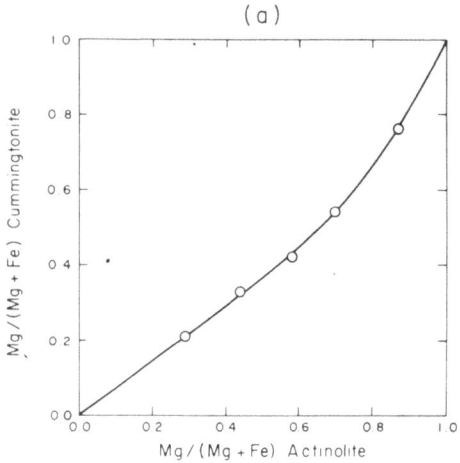

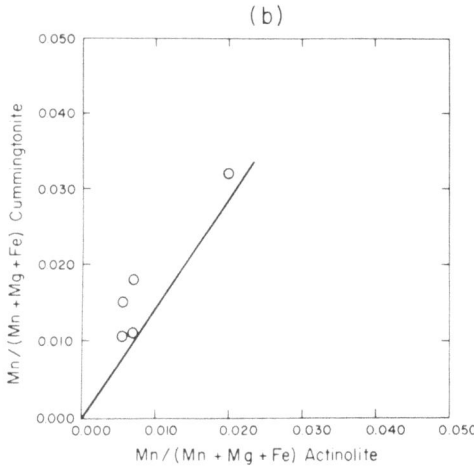

Figure 14

Fractionation of manganese, iron and magnesium between coexisting actinolites and cummingtonites, metamorphosed iron formation, Quebec (Mueller, 1960). The major element distribution shown in (a) approximates mass action law behavior; minor element distribution shown in (b) obeys the infinite dilution law.

tinction of ferrous and ferric iron. Accordingly exchange equilibria considered involve total iron contents rather than either Fe^{+2} or Fe^{+3}. Relations are presented in Fig. 14. The curve shown in Fig. 14b was computed from the manganese fractionation between actinolite + calcic pyroxene and between cummingtonite + calcic pyroxene because of the scatter of the two-amphibole data; this calculation involves the errors of both curves, as stated by MUELLER (1960, p. 475). From Figs. 14a and 14b it may be noted that $Fe/(Mg+Fe)$ and $Mn/(Mg+Fe+Mn)$ ratios for cummingtonite are slightly larger than those for coexisting actinolite*; this phenomenon presumably results because the largest ions available, iron and manganese, are strongly concentrated in the M_4 position where calcium is not abundant. This deduction is supported by the recent structural refinement of cummingtonite by FISCHER (1966), and by MÖSSBAUER and infra-red studies of Fe-Mg clinoamphiboles by BANCROFT et al. (1967). Ordering of cations among octahedral sites in both cummingtonite-grunerite and tremolite-actinolite series (BURNS, 1966; BURNS and STRENS, 1966) means that exchange equilibria involving these amphiboles cannot be explained adequately based on ideal solution theory models.

Unpublished work by the present author dealing with coexisting glaucophanes + actinolites reveals somewhat similar relationships to those elucidated by MUELLER (1960, 1962). Analyses were performed using an electron microprobe, so as with MUELLER's study, distinction between ferrous and ferric iron was not practicable. Preliminary results from schists of the outer metamorphic belt of Japan and the California Coast Ranges are compared in Fig. 15. Sodic amphiboles concentrate iron with respect to coexisting calcic amphiboles as shown in Fig. 15a; this phenomenon probably means that M_2 is occupied preferentially by Fe^{+3} (and Al) ions in glaucophane and by Mg in actinolite, hence the $Mg/(Mg+Fe)$ ratio of actinolite exceeds that of glaucophane. Japanese coexisting pairs exhibit a less pronounced fractionation of iron and magnesium than Californian counterparts. These relations are in agreement with the higher temperature crystallization of the Japanese schists compared to the Californian rocks postulated by ERNST and SEKI (1967), based on comparison of mineral assemblages with experimentally determined phase equilibria. Manganese is concentrated in calcic amphiboles relative to sodic amphiboles but, as indicated in Fig. 15b, the data show considerable scatter. Apparently the M_4 site in actinolite is slightly

---

*This observation has been documented recently by KLEIN (*in press*) who presents new electron microprobe analyses of 37 amphibole pairs. In general, anthophyllites, like cummingtonites, concentrate Mn and Fe with respect to coexisting calcic amphiboles which preferentially incorporate magnesium.

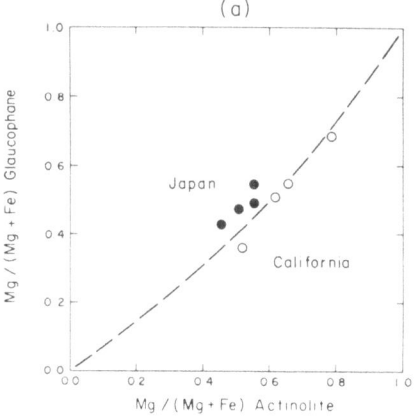

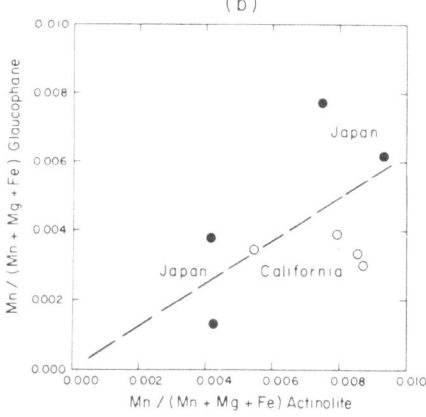

Figure 15

Fractionation of manganese, iron and magnesium between coexisting actinolites and glaucophanes, outer metamorphic belt of Japan and California Coast Ranges. The major element distribution shown in (a) approximates mass action law behavior; minor element distribution shown in (b) crudely obeys the infinite dilution law.

larger than the equivalent position in glaucophane, so concentrates the manganese; however, the situation is complicated by the fact that some of the Mn probably resides in $M_1 + M_3$ positions in both amphiboles.*

---

*KLEIN (*in press*) demonstrated that for the single amphibole pair, manganoan cummingtonite + magnesioriebeckite, Mn is markedly enriched in the Fe-Mg amphibole. Sodic amphiboles evidently do not incorporate large amounts of manganese if other double-chain silicates are present.

Chapter IV.

HYDROTHERMAL PHASE EQUILIBRATION
AND NATURAL STABILITY

In previous chapters we have considered the chemical range of amphibole compositions, and the structural basis for this variation. The next three chapters will summarize experimentally determined stability relationships involving double-chain silicates; an attempt also will be made to apply these data to appropriate natural occurrences. Hydrothermal methods and equilibration techniques are discussed briefly in this section, as well as disparities between natural conditions and experimentally controlled conditions.

Experimental Techniques

Starting materials for most laboratory studies have consisted of stoichiometric proportions of the oxides equivalent to the amphibole bulk composition. A small excess of water is added to the charge to insure complete H_2O saturation of the resultant synthetic assemblage, as reflected by the presence of an aqueous fluid phase. Other techniques involve the preparation of gels, glass or oxalate mixtures of the correct bulk composition ($+$ excess $H_2O \pm CO_2$). In any case, such starting materials are chosen for experimental investigations because of their great reactivity Another approach involves (1) the crystallization of the initial oxides to a mixture of relatively less reactive synthetic stable and metastable phases or (2) intimately mixing pure, analyzed natural minerals of both stable and metastable associations; either of these mixtures of phases is then employed as the charge.

The starting material is sealed in an inert metal container of negligible strength such as silver, gold, platinum or silver-palladium alloy, which in turn is then placed in an externally heated pressure vessel or "bomb." The pressure vessel is pumped to the desired value, the customary pressure medium being H_2O, although argon or other noncorrosive gases are employed occasionally, and heated by placing it in a resistance-furnace. The excess H_2O within the charge container expands or contracts at elevated temperatures and pressures until the total pressure within the vessel and the sample capsule are equal; thus P_{fluid} within the charge is equal to P_{total}. Because the aqueous phase is supercritical the term fluid is used rather than gas or vapor. This fluid is largely H_2O, but is diluted by solution of other constituents from the starting material, especially alkalis where present, and silica; if oxalate mixtures are employed, CO_2 is an important contaminant. Hence, in some experiments, P_{H_2O} may be somewhat less than P_{fluid}. Two variables are thus involved here, both of which affect amphibole stability relationships: (1) total pressure; and (2) partial pressure of H_2O, or more correctly, the fugacity or activity of H_2O. A typical sample capsule + bomb + furnace assembly is illustrated in Fig. 16. For details regarding pressure vessel design and fabrication, see TUTTLE, 1948, 1949, and LUTH and TUTTLE, 1963. At the conclusion of the experiment, the pressure vessel is removed from the furnace, quenched, and the condensed run products examined by conventional petrographic and x-ray methods.

Where the starting material contains one or more elements of variable valence state, such as carbon, iron, nickel, copper or manganese, the physical variables which influence phase relationships include oxygen fugacity as well as pressure and temperature. The buffer technique of EUGSTER (1957, 1959) has been employed routinely to define values of f_{O_2}. This method requires that the charge capsule (a container consisting of a metal permeable to hydrogen) be surrounded by an oxygen buffer assemblage $+H_2O$ in an outer capsule (the container metal being nearly impermeable to H_2). Equilibration of the oxygen buffer assemblage and fluid defines fugacities of hydrogen and oxygen in the outer capsule; through diffusion of the H_2, the f_{H_2} and hence the f_{O_2} of the inner capsule are controlled through the dissociation of H_2O. An oxygen fugacity

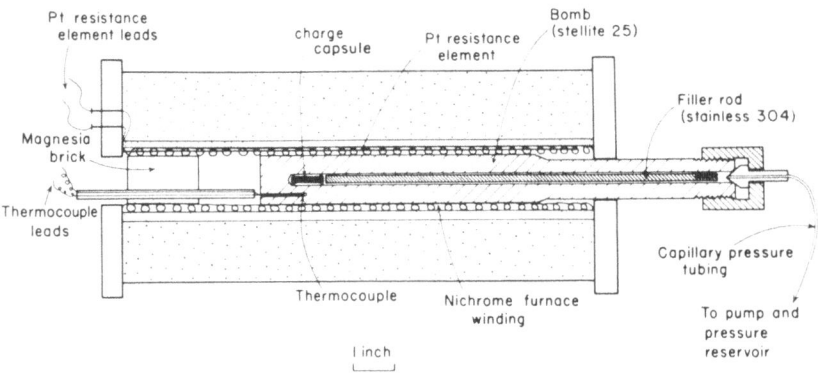

Figure 16

Typical hydrothermal synthesis assembly, showing furnace, pressure vessel and sample container (through the courtesy of F. R. Boyd).

versus temperature plot for the various buffer assemblages is presented in Fig. 17. For a general discussion of oxygen buffers, see EUGSTER and WONES, 1962. Variation in total pressure does not significantly affect the value of log f_{O_2} defined by a specific buffer assemblage, at constant temperature.

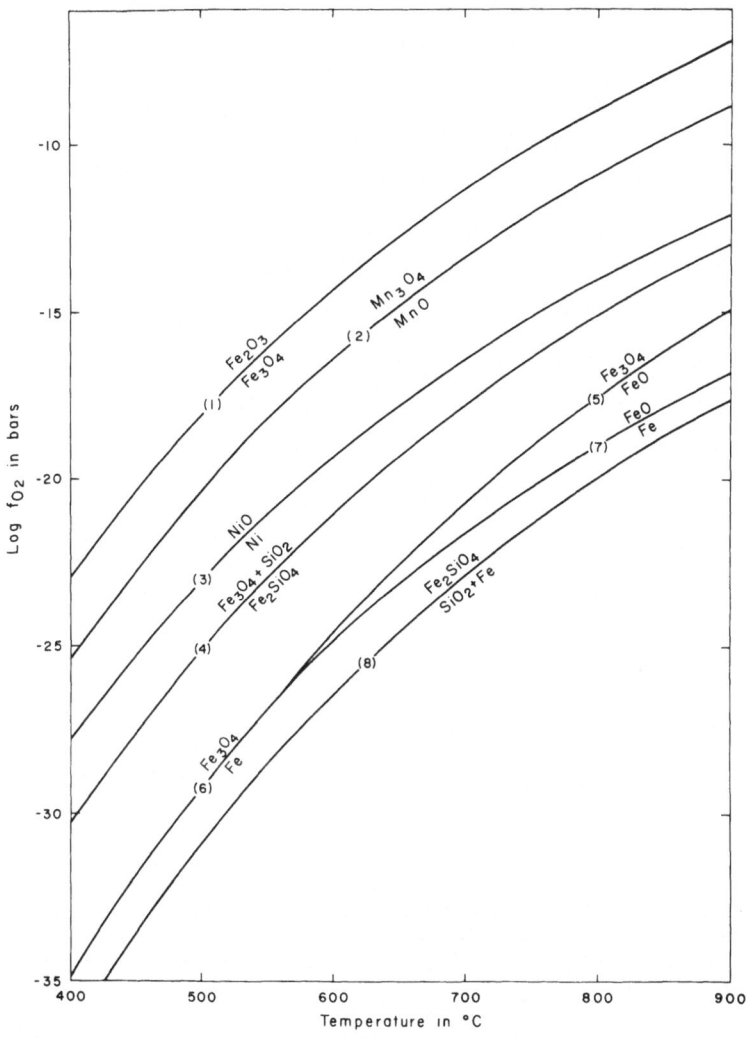

Figure 17

Log f_{O_2}-T plot for oxygen buffer assemblages; the influence of total pressure is negligible, as discussed by Ernst (1960) and Wones and Eugster (1962).

Demonstration of Experimental Equilibria

Many synthetic amphibole reactions are exceedingly sluggish, and in some cases it is difficult to establish whether or not equilibrium has been achieved, or even closely approached. For a given bulk composition the equilibrium state can be defined as the one giving the phase assemblage with the lowest possible Gibbs free energy. All other assemblages are thus unstable. It follows for an equilibrium reaction $A = B + C$ that, if A represents the minimum Gibbs free energy configuration in a specific P-T stability region, $B + C$ will constitute the lowest Gibbs free energy assemblage in an adjoining P-T stability region. In this case the two assemblages are related by a univariant P-T curve, along which both have the same Gibbs free energy.

As previously mentioned, mixtures of oxalates, oxides, gels or glasses are highly reactive. Unfortunately, due to the elevated Gibbs free energy of such materials, the initial charge may be transformed readily into a metastable but persistent crystalline assemblage, as shown schematically in Fig. 18a. This latter phase association has a Gibbs free energy lower

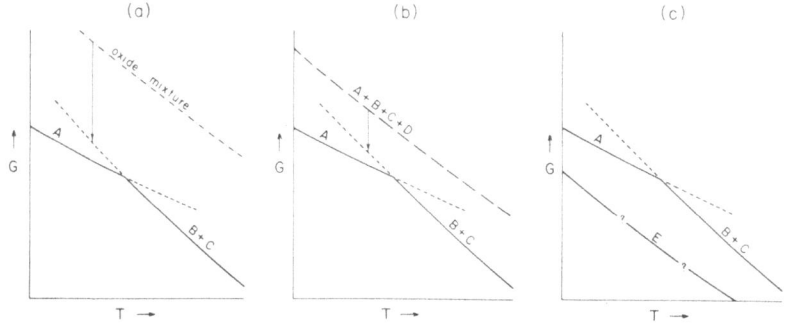

Figure 18

Isobaric Gibbs free energy-temperature diagrams for equilibrium versus meta-stable equilibrium. Remember that $(\partial G/\partial T)_p = -S$, so because all assemblages have positive entropy, curve slopes are negative and steeper slopes characterize relatively higher entropy assemblages. Metastable curve extensions are shown as short dashes. (a) High energy oxide mixture spontaneously reacts (arrow) and *may* form metastable assemblage $B + C$ in the stability field of A; eventually $B + C$ will spontaneously react to form the stable assemblage. (b) Metastable assemblage $A + B + C + D$ spontaneously reacts (arrow) and *may* form the lower energy but metastable configuration $B + C$; again, $B + C$ will spontaneously react to form the stable assemblage of A. (c) Hypothetical configuration E, of lower Gibbs free energy than either A or $B + C$; if assemblage E is never produced experimentally, metastable equilibrium only will have been demonstrated, and we must rely on observed natural phase relations for clues indicating the more stable assemblages.

than that of the starting material, but in excess of the stable configuration; it may persist indefinitely, thus impeding determination of the stable assemblage (for general discussions see FYFE, 1960 and MACKENZIE, 1965).

The alternative experimental approach of using crystalline mixtures of stable and metastable phases, and observing growth to indicate the equilibrium assemblages must be considered carefully. The critical point involved here is this: regarding the reaction A = B + C under physical conditions where A is stable, one must be certain that the initial charge consists only of phases A, B and C. If D, having a very high Gibbs free energy is also present, it may react to yield metastable B + C of lower Gibbs free energy than D. However, A has the lowest Gibbs free energy of all. Therefore, as shown diagrammatically in Fig. 18b, the increase of B + C in the experimental run products would lead to the erroneous conclusion that the assemblage B + C was stable with respect to A.

Experimental demonstration of equilibrium thus requires that each of two assemblages, related by a specific reaction, be shown to grow from one another within their individual stability fields; this proof is known as reversal of reaction. Again considering the reaction A = B + C, one could conceive of an hypothetical configuration E, which would have a lower Gibbs free energy than either A or B + C over the P-T range considered, as shown in Fig. 18c; but if it cannot be synthesized, we have no information regarding physical conditions of stability. However, if E occurs in nature under P-T conditions known to approximate those of the laboratory synthesis, then the reaction A = B + C represents metastable equilibrium, not stable equilibrium.

In summary, the minimum or stable energy configuration for any specific system cannot be proven; what experimental (and observational) petrologists strive to demonstrate is equilibrium with respect to the phase assemblages studied.

Contrasts in Natural and Synthetic Conditions

Almost all experimental determinations of amphibole phase relationships have been carried out utilizing starting materials in stoichiometric proportions to form the pure phase—either as an end-member, or as members of a binary solid solution series. Moreover, most laboratory syntheses have been performed under conditions where P_{H_2O} is maximized; that is, the departure of P_{H_2O} from $P_{total} = P_{fluid}$ arises solely because of contamination of the aqueous phase by constituents present in the amphibole itself. With few exceptions rocks have bulk compositions which deviate to some extent from that of constituent amphiboles. Furthermore, in some cases a separate fluid phase may not have been present during amphibole crystallization; in certain other instances, although fluid apparently was present, it may have been diluted by components such as CO_2 not present in amphiboles.

These contrasts between natural and experimental conditions are such that in nature, the P-T stability ranges of amphiboles are more severely restricted than evident from laboratory investigations. The explanation is illustrated schematically in Fig. 19.

At constant total pressure, and with $P_{fluid} = P_{total}$, the temperature T_1 for the decomposition of tremolite for instance, to form 2 diopside + 3 enstatite + quartz + fluid exceeds T_2, the temperature at which tremolite reacts with forsterite to produce 2 diopside + 5 enstatite + fluid. The phenomenon reflects the fact that forsterite and quartz are incompatible, and spontaneously react to form enstatite with a relative decrease in Gibbs free energy for the high temperature assemblage equivalent in bulk composition to the low temperature assemblage; this in turn causes G-T curves for the amphibole-bearing and anhydrous condensed assemblages to intersect at a lower temperature. Thus, as illustrated in Fig. 19a, departure of rock bulk composition from that of the amphibole itself reduces the thermal stability limit of the amphibole. On the other

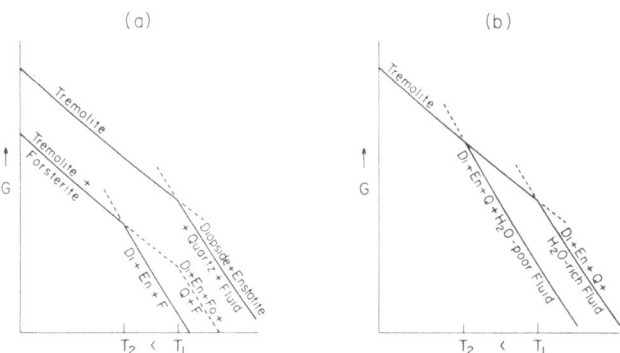

Figure 19

Isobaric Gibbs free energy-temperature diagrams illustrating the influence of changing composition of the system on the thermal stability limit of a pure phase. Metastable curve extensions are shown as short dashes. T_1 = isobaric high temperature stability limit of tremolite in a rock of its own bulk composition, T_2 = isobaric high temperature stability limit of tremolite in a system (a) undersaturated with respect to silica and (b) with fluid phase empoverished in H_2O. Abbreviations include: Di = diopside; En = enstatite; Q = quartz; Fo = forsterite; F = fluid.

hand, the isobaric decomposition of tremolite in the presence of excess quartz takes place at temperature T_1, because both low and high temperature assemblages are saturated with respect to silica. In general, where addition of another constituent promotes a spontaneous reaction, the stability field of the newly produced assemblage expands at the expense of the pure phase, provided the pure phase is compatible with the added species. For other discussions of the effect, see ROSENFELD, 1961; FYFE et al., 1961.

In Fig. 19b, which is also an isobaric total pressure diagram, we see that a similar effect results from dilution of the aqueous fluid due to the negative value for the Gibbs free energy of mixing; the reduced activity of H_2O is reflected in diminished Gibbs free energy for the high temperature, fluid-containing assemblage. Thus decreased activity, or fugacity, of H_2O also lowers the thermal stability limit of the amphibole ($T_2 < T_1$), and for all other hydrous minerals as well. For a general discussion of the role of fluids in crystallization, the reader is referred to papers by: YODER, 1952; THOMPSON, 1955; GREENWOOD, 1961; ZEN, 1961; and BARNES and ERNST, 1963.

For reactions in certain metamorphic terranes, dilution of the aqueous phase by CO_2 seems to be the most plausible explanation of lowered activity, or fugacity of H_2O. For example, WONES and EUGSTER (1965, p. 1267) concluded that Adirondack gneisses and amphibolites probably crystallized at f_{H_2O} values of 0.1–10 bars, several orders of magnitude less than the lithostatic pressure; WONES and EUGSTER arrived at these deductions based on experimentally determined phase relationships for the biotites and for tremolite. GREENWOOD (1963) studied mineral parageneses in calc silicate rocks from the Balmat area of the Adirondacks as well as in serpentinites from northeast Maryland and southeast Pennsylvania, and came to similar conclusions regarding activity of H_2O judging from stability relationships of anthophyllite. In many cases, dilution of the aqueous phase probably resulted from the progressive decarbonation of carbonate rocks (BOWEN, 1940), or due to the oxidation of graphite. For a general discussion of this latter problem, see: MIYASHIRO, 1964; FRENCH and EUGSTER, 1965; FRENCH, 1966.

Thus, we have seen that, where demonstration of equilibrium has been successful, laboratory syntheses of amphiboles yield information regarding the *maximum* P-T range for these minerals in nature. Departure of rock bulk compositions from those investigated hydrothermally, and variations of f_{H_2O} (and in some cases, f_{O_2}) restrict the physical conditions of amphibole crystallization in rocks.

Chapter V.

EXPERIMENTAL PHASE RELATIONS AND OCCURRENCE OF THE IRON-MAGNESIUM AMPHIBOLES

Anthophyllites, $\circ(Mg,Fe^{+2})_7Si_8O_{22}(OH)_2$

BOWEN and TUTTLE (1949) first directed the attention of experimental petrologists to anthophyllite by producing it as a metastable breakdown product of talc; these authors were unable to demonstrate a stability field for anthophyllite and suggested that its production might depend on the absence of H_2O-rich fluid. YODER (1952) experienced similar results in his investigation of the system MgO-Al_2O_3-SiO_2-H_2O, and advanced the concept of the "water deficient region," that is, chemographic space in which a stable fluid phase is prohibited; YODER suggested that anthophyllite might be confined to such a compositional region, at least in the investigated system.

More recently P-T stability relations for anthophyllite, under conditions where $P_{fluid} = P_{total} \approx P_{H_2O}$ have been determined by GREENWOOD (1962b, 1963) employing starting materials consisting of synthetic crystals; in addition, FYFE (1962) showed that natural anthophyllite, used as starting material, is stable in the presence of an aqueous fluid. Both of these investigators demonstrated that equilibrium had been attained by reversing the reactions. The low temperature stability limit, high temperature stability limit, and intermediate ternary reaction relations involving the Mg end-member are shown in Figs. 20 and 21, taken from work by GREENWOOD (1963). At moderate fluid pressures, anthophyllite has a thermal stability range of approximately 80 C° in a system of its own bulk composition, i.e., at 1000 bars P_{fluid}, this amphibole is stable between 667–745° C. As indicated in Fig. 20, at low temperatures 9 talc + 4 forsterite react to form 5 anthophyllite + fluid along curve 1; with further elevation of temperature, anthophyllite decomposes along curve 4 to give 7 enstatite + quartz + fluid. The reaction anthophyllite = talc + 4 enstatite involves a volume decrease, hence GREENWOOD was able to use calorimetric data to calculate the high pressure stability limit of approximately 20–25 kb for $\circ Mg_7Si_8O_{22}(OH)_2$ illustrated in Fig. 21.

GREENWOOD also showed that anthophyllite is an intermediate metastable product of the decomposition of talc at high temperatures.

Disintegration of the sheet silicate apparently provides double-chain strips which in turn promote nucleation of the amphibole at temperatures above its own stability field. From a rate study GREENWOOD (1963, Fig. 5, Table 9) demonstrated that, although growth of anthophyllite is rapid initially, it eventually is replaced by the stable condensed assemblage of enstatite + quartz.

Syntheses of intermediate members of the anthophyllite-ferroanthophyllite series have been reported by HELLNER et al. (1965), and by HINRICHSEN (1966). These authors employed mixtures of oxalates, oxides and hydroxides, so the fluid phase contained CO_2 as well as H_2O. At

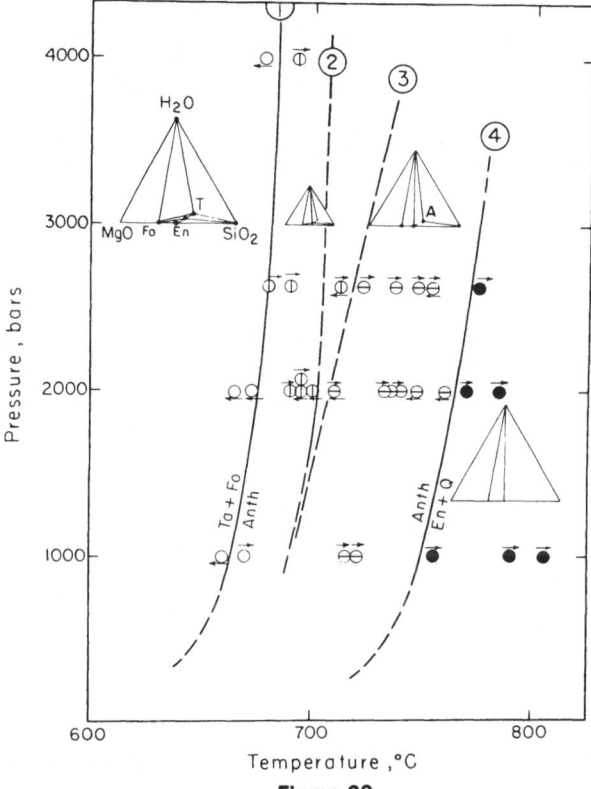

Figure 20

Experimentally determined phase equilibria for the system MgO-SiO_2-H_2O under conditions where $P_{fluid} = P_{total}$ (Greenwood, 1963). Arrows indicate direction from which equilibrium was approached. Curve (1) represents the low temperature stability limit of anthophyllite, curve (4) the high temperature stability limit. Abbreviations are as follows: F = Fo = forsterite; T = Ta = talc; E = En = enstatite; A = Anth = anthophyllite; Q = quartz.

1000 bars H_2O pressure, HELLNER *et al.* (1965) computed values for P_{CO_2} on the order of 50–100 bars for the investigated conditions; they used sealed gold charge containers which were impermeable to hydrogen. HINRICHSEN (1966) employed sealed double capsules and the magnetite-iron and magnetite-wüstite oxygen buffers. For all these experiments, equilibration between the fluid phase and the contrasting charge assemblages characteristic of different temperatures (and also due to reaction with the buffer where used) would be expected to influence the fugacities

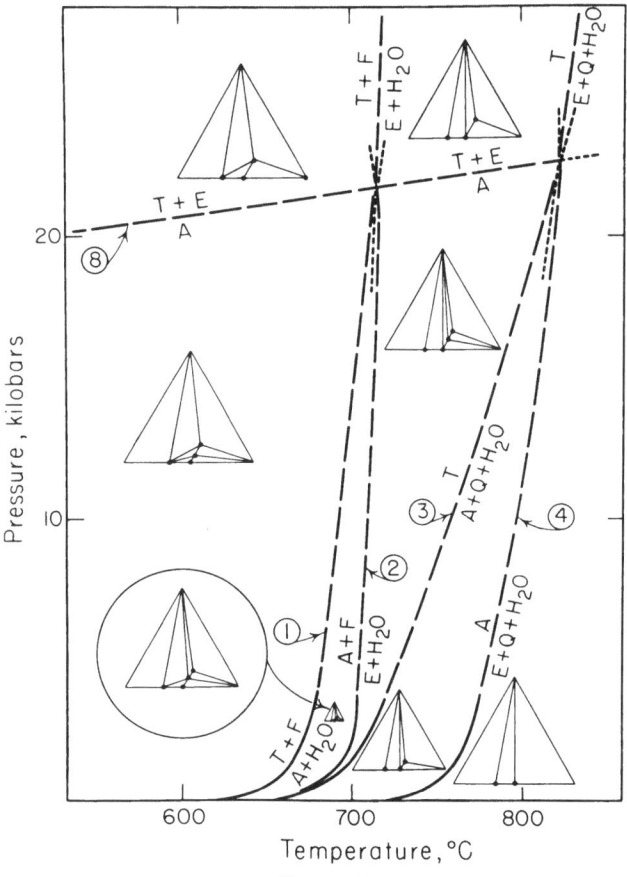

Figure 21

Schematic P_{fluid}-T diagram for the system $MgO \cdot SiO_2 \cdot H_2O$ (Greenwood, 1963). Curves are solid where experimentally verified, and are dashed where projected or calculated (short dashes represent metastable extensions). Location of the high pressure stability limit of anthophyllite is only approximate. Abbreviations are the same as in Fig. 20.

of both CO_2 and H_2O. In the earlier study by HELLNER *et al.*, equilibrium
was not demonstrated, but in the later, which represents a revision of a
portion of the earlier work, reactions were reversed; results were checked
employing the ion exchange method of ORVILLE (1963). HINRICHSEN's
results, presented in Fig. 22, indicate that synthetic solid solution between $^\circ Mg_7Si_8O_{22}(OH)_2$ and $^\circ Fe_7^{+2}Si_8O_{22}(OH)_2$ extends to approximately 60 percent of the iron end-member. For more iron-rich bulk compositions, orthoamphibole is not stable. Similar to the phase relationships
for anthophyllite (GREENWOOD, 1963), intermediate members of this
series have a thermal range of approximately 100 C° or less.

CHOUDHURI and WINKLER (1967) synthesized anthophyllite + hornblende assemblages from an initial charge consisting of natural chlorite,
talc, tremolite and quartz; the ferromagnesian components were present
in the ratio $Mg_{84}Fe_{16}$ in the starting material. At 1000 bars fluid pressure, hornblende + anthophyllite appeared over the temperature interval
550–715° C. The hornblende is more refractory than the orthoamphibole
and persists to temperatures exceeding the anthophyllite breakdown
curve.

Phase relations for anthophyllite coexisting with a $CO_2 + H_2O$-bearing
fluid phase have been calculated recently by GREENWOOD (1967, Figs.
4, 8). At low temperatures and high CO_2 concentrations, the assemblage
magnesite + quartz replaces the amphibole. Therefore the low tempera-

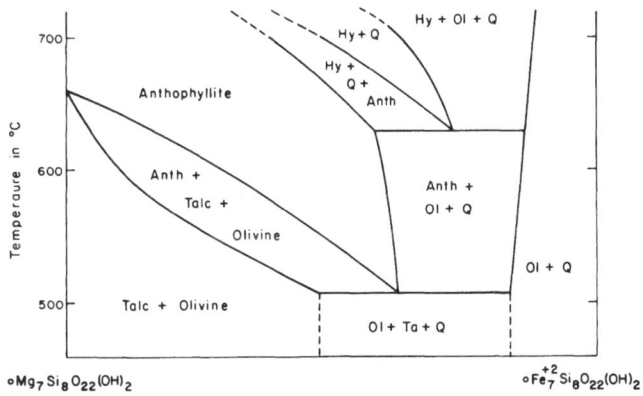

Figure 22

Experimentally determined isobaric T-x diagram for the pseudo-binary join anthophyllite-ferroanthophyllite (Hinrichsen, 1966). Values for P_{H_2O} and P_{CO_2} are
1000 and about 50 bars respectively; oxygen fugacities are defined by the magnetite-iron and magnetite-wüstite buffers. Abbreviations are: Anth = anthophyllite;
Ol = olivine; Ta = talc; Q = quartz; Hy = hypersthene. All assemblages coexist
with fluid.

ture stability limit of anthophyllite cannot be reduced indefinitely through dilution of the fluid phase by a chemically active species; instead another assemblage of equivalent bulk composition is produced.

Gedrites, $\circ(Mg,Fe^{+2})_5Al_2Si_6Al_2O_{22}(OH)_2$

Starting with natural materials (chlorite + quartz + ferrogedrite seeds) of magnesium-iron bulk proportions $Mg_{56}Fe_{44}$, AKELLA and WINKLER (1966) determined the high temperature and low temperature P-T stability limits of the aluminous orthoamphibole gedrite, in the presence of cordierite + quartz. The initial natural ferrogedrite employed as seed material contained 2.00 weight percent Na_2O. Reactions were reversed. The amphibole formed was deduced to contain abundant aluminum and ferromagnesian constituents in the proportions $Mg_{65-70}Fe_{35-30}$. Oxygen fugacities were not controlled, but apparently approximated those defined by the magnetite + quartz − fayalite buffer due to equilibration between the H_2O pressure medium and the bomb walls. (Gold is practically impermeable to hydrogen, but on long runs diffusion of H_2 results in hydrogen-oxygen equilibration between charge and pressure medium if buffers are not employed.) Because the reacting phases are intermediate solid solutions, the univariant curves presented by AKELLA and WINKLER shown in Fig. 23 actually must be P-T zones over which reactants and

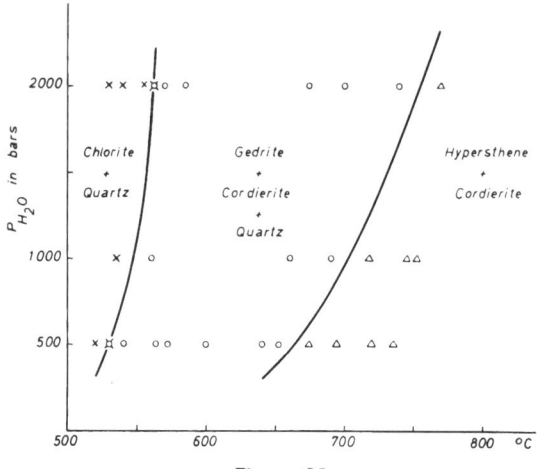

Figure 23

Experimentally determined P_{fluid}-T curves for the reactions chlorite + quartz = gedrite + cordierite + H_2O, and gedrite + quartz = hypersthene + cordierite + H_2O, employing natural starting materials of atomic proportions $Mg_{56}Fe_{44}$ (Akella and Winkler, 1966). All assemblages are in equilibrium with a fluid phase.

products coexist in varying proportions and with changing Fe^{+2}-Mg ratios of the phases; judging from their data, these zones are probably at least 10–15 C° wide. Intermediate gedrites have a thermal stability range of approximately 155 C° at moderate pressures; i.e., at 1000 bars P_{fluid}, this amphibole is stable in association with cordierite + quartz between 548–704° C. The presence of iron lowers both high and low temperature stability limits of Al-rich amphibole with respect to those of anthophyllite.

In contrast to the work of AKELLA and WINKLER (1966), HINRICHSEN (1966) was unable to produce gedrites using charges of bulk compositions within the system MgO-FeO-Al_2O_3-SiO_2-CO_2-H_2O. Aluminous ortho-amphiboles were synthesized successfully only in the presence of small amounts of sodium, 2 mole percent; demonstration of equilibrium was claimed. The oxygen buffers iron-magnetite and magnetite-wüstite were employed. The low temperature limit for intermediate Fe-Mg gedrite solid solutions determined by HINRICHSEN lies about 150 C° above the value published by AKELLA and WINKLER, at approximately 1000 bars H_2O pressure. The low temperature assemblage equivalent in bulk composition to gedrite as stated by HINRICHSEN is cordierite + olivine + spinel (+ fluid), an anomalous low grade compatibility considering the fact that the condensed assemblage is anhydrous. AKELLA and WINKLER investigated a silica excess bulk composition, but the presence of additional SiO_2 in HINRICHSEN's experiments would have raised his low temperature limit for gedrite due to the reaction of spinel and olivine with silica to produce cordierite + hypersthene, thus increasing the disparity between the two works. However, if the synthetic gedrites contain large amounts of ferric iron, differences in oxygen fugacity ranges for the two sets of experiments might account for part of the discrepancy.

Cummingtonite-Grunerite, $\circ(Mg,Fe^{+2})_7Si_8O_{22}(OH)_2$

Synthetic members of the fluorcummingtonite-fluorgrunerite series were prepared in sealed, evacuated silica-glass tubes by BOWEN and SCHAIRER (1935) in their study of the system $MgSiO_3$-$FeSiO_3$-NaF. Equilibrium was not demonstrated, nor did these authors prove that the synthetic amphiboles lacked sodium. In their investigations in the system Fe-Si-O-H, FLASCHEN and OSBORN (1957) and SMITH (1957) were unable to locate a stability field for grunerite, although the former authors reported amphibole as a metastable decomposition product of minnesotaite.

SCHÜRMANN (1966) succeeded in synthesizing a series of iron-magnesium clinoamphiboles from cummingtonite of the ferromagnesian proportions $Mg_{65}Fe_{35}$ to the iron end-member, but only for iron oxalate-bearing bulk compositions which contained 2.7 weight percent CaO. Demonstration of equilibrium was claimed. Calcium-bearing grunerite was synthesized between 450° C and 595° C at 1000 bars H_2O pressure + 80 bars CO_2 pressure. Intermediate Ca-bearing cummingtonite-grunerites were obtained at higher temperatures. The low temperature assemblages of equivalent bulk composition are thought to be talc + olivine, talc + olivine + quartz, and for iron-rich bulk compositions, olivine + quartz (all + fluid). The high temperature assemblage also consists of olivine + quartz for bulk compositions near the grunerite end-member, which evidently necessitates a pair of intersections for isobaric G-T curves representing the two assemblages. Although SCHÜRMANN (1966) suggests that CaO and/or MnO may be essential for the production of cummingtonite in nature, many of the chemical analyses of this mineral group presented by DEER et al. (1963, Table 36) and all nine of those of KLEIN (1964, Table 2) indicate negligible CaO, and some contain negligible CaO + MnO. Furthermore, as mentioned in the discussion of Fig. 13, certain orthoamphiboles contain similar amounts of lime (and/or soda).

Natural Occurrences of the Iron-Magnesium Amphiboles

The iron-magnesium amphiboles are virtually restricted to metamorphic rocks and are the least abundant of the three groups of amphiboles. Cummingtonite-bearing hornblende gabbros do occur in the Abukuma Plateau, Japan, and in Finland according to SHIDO (1958, p. 196–197) and SEITSAARI (1952). WARREN (1903) reported grunerite from a fayalite-bearing pegmatite, Rockport, Massachusetts; however, textural relations suggest that this amphibole may represent deuteric or hydrothermal reaction among quartz, fayalite and aqueous fluid. Judging from the experimentally determined phase relationships, cummingtonite-grunerites are stable only at submagmatic temperatures for mafic igneous rocks.

Members of the cummingtonite-grunerite series occur in metamorphosed iron formations. MUELLER (1960), KRANCK (1961) and KLEIN (1966) reported the coexistence of actinolite + cummingtonite-grunerite from east-central Quebec, and GUNDERSON and SCHWARTZ (1962, p. 85, 123) described the assemblage hypersthene + hedenbergite + cummingtonite-grunerite ± fayalite, as well as the coexistence of cummingtonite-grunerite with hornblende or actinolite. Higher grade rocks contain abundant fayalite, lower grade, or retrograde rocks contain more hydrous

Fe-Mg silicates such as minnesotaite, stilpnomelane and nontronite. Cummingtonite-grunerite also appears in the middle grades of metamorphosed mafic igneous rocks (*e.g.*, Finland, ESKOLA and KERVINEN, 1936; Japan, SHIDO, 1958). A typical paragenesis is shown in Fig. 24. SHIDO explained the prevalence of cummingtonite in high grade, relatively low pressure terranes and its absence from high pressure terranes (for an introduction to metamorphic belts of contrasting P-T regime, see MIYASHIRO, 1961) by the following reaction, which involves a volume decrease of approximately 10 percent:

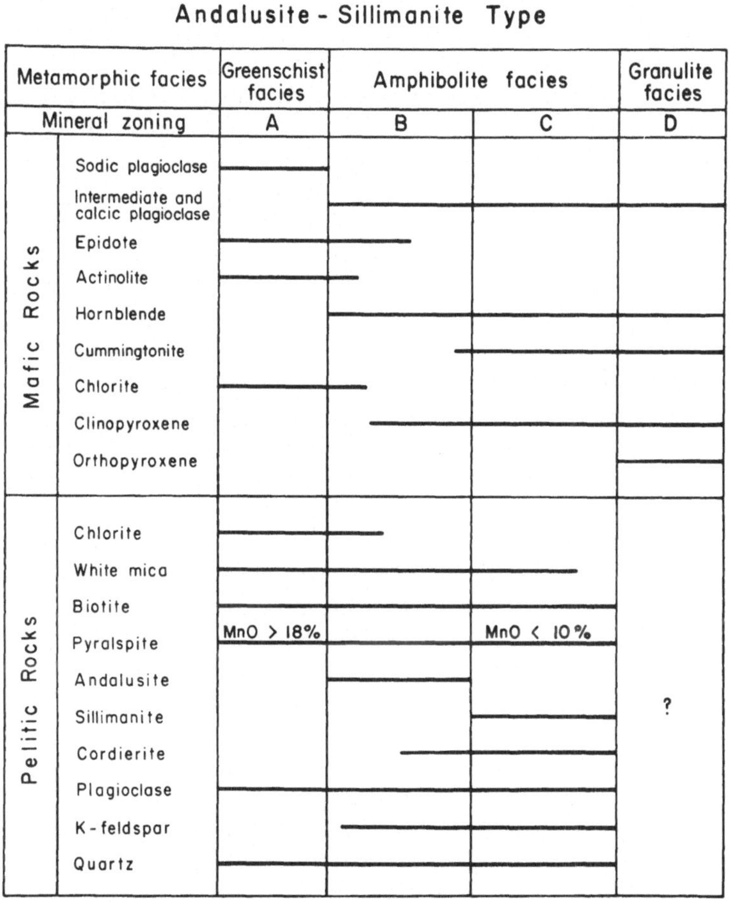

Andalusite - Sillimanite Type

Metamorphic facies		Greenschist facies	Amphibolite facies		Granulite facies
Mineral zoning		A	B	C	D
Mafic Rocks	Sodic plagioclase				
	Intermediate and calcic plagioclase				
	Epidote				
	Actinolite				
	Hornblende				
	Cummingtonite				
	Chlorite				
	Clinopyroxene				
	Orthopyroxene				
Pelitic Rocks	Chlorite				
	White mica				
	Biotite				
	Pyralspite	MnO > 18%		MnO < 10%	
	Andalusite				
	Sillimanite				?
	Cordierite				
	Plagioclase				
	K - feldspar				
	Quartz				

Figure 24

Mineral parageneses in progressive metamorphism of the central Abukuma Plateau, Japan (Miyashiro, 1958, 1961; Shido, 1958).

14 anorthite$_{ss}$ + 3 cummingtonite + 4H$_2$O $\xrightarrow{\text{incr.P}}$ 7 tschermakite$_{ss}$ + 10 quartz.

A parallel reaction involving the Ab component of plagioclase also could be written:

14 albite$_{ss}$ + 3 cummingtonite + 4H$_2$O $\xrightarrow{\text{incr.P}}$ 7 glaucophane$_{ss}$ + 10 quartz.*

In general, iron-magnesium amphiboles react with the CaAl$_2$Si$_2$O$_8$ and NaAlSi$_3$O$_8$ components of plagioclase to produce more Al-rich hornblendes and more sodic amphiboles in the relatively high pressure terranes. Of course, the displacement of this reaction depends critically on the activities of all participating species, hence elevated H$_2$O fugacities and high normative plagioclase contents of certain rocks tend to favor production of aluminous amphiboles.

Members of the anthophyllite-gedrite group are accompanied typically by cordierite and commonly sillimanite, rather than garnet in high rank amphibolites and gneisses (Eskola, 1914); this association suggests relatively elevated temperatures and moderate pressures. However, coexistence of gedrite and kyanite from Idaho (Hietanen, 1959) indicates that the aluminous iron-magnesium amphibole may represent a higher pressure equivalent of the anthophyllite + cordierite + sillimanite association. The reaction of chlorite + quartz → gedrite + cordierite + H$_2$O (Fig. 23) has been correlated by Akella and Winkler (1966) with the lower P-T boundary of the hornblende hornfels facies under conditions where fluid pressure equals total pressure. Based on other reactions such as 6 muscovite + 2 biotite + 15 quartz = 3 cordierite + 8 K-feldspar + 8 H$_2$O (Winkler, 1965, p. 58), these authors believe that the pyroxene hornfels facies commences at lower temperatures than the maximum thermal stability limit of gedrite shown in this figure; the reaction shown in Fig. 23 involves conversion of gedrite + quartz to hypersthene + cordierite + H$_2$O.

Anthophyllite occurs in metamorphosed mafic igneous rocks and in hornfels, but is characteristic of siliceous magnesian marbles and altered ultramafic rocks. Employing the phase diagram presented for $^\circ$Mg$_7$Si$_8$O$_{22}$(OH)$_2$ in Fig. 20, and the estimated physical conditions of approximately 500° C, several kilobars lithostatic pressure estimated by Engel and Engel (1960) and Doe (1962), Greenwood (1963) demonstrated that the equilibrium pressure of H$_2$O must have been on the order of a few bars for Adirondack anthophyllite; apparently if fluids were present, they were CO$_2$-rich due to devolatilization of the enclosing carbonate

---

*In both reactions, ss denotes solid solution, hence albite$_{ss}$ refers to the component Ab in plagioclase solid solution.

layers. Regarding serpentine-talc-anthophyllite parageneses, GREENWOOD showed that the equilibration of these phases depends on very low equilibrium pressures of H_2O, and a pronounced pressure or temperature gradient.

Although Na_2O and CaO need not be essential components in cummingtonite-grunerite and in anthophyllite-gedrite, experimental syntheses by SCHÜRMANN (1966) and HINRICHSEN (1966) suggest that sodium and calcium ions do enter these structures and possibly extend their thermal stability ranges. Coexisting ortho- and clinoamphiboles of the Fe-Mg group would be expected to show different concentrations of these components, calcium presumably being enriched in the cummingtonite-grunerite series, and sodium in the anthophyllite-gedrite series.

Chapter VI.

EXPERIMENTAL PHASE RELATIONS AND OCCURRENCE OF THE CALCIC AMPHIBOLES

Tremolite, $\square Ca_2Mg_5Si_8O_{22}(OH)_2$

The high temperature stability relationships of tremolite were determined by BOYD (1959). As is common in most other synthetic hydrothermal investigations, $P_{fluid} = P_{total}$. The starting material in this study consisted of roughly equal proportions of synthetic tremolite and the synthetic high temperature assemblage of equivalent bulk composition, enstatite + diopside + quartz + fluid; stability fields were deduced on the basis of recognizing the contrasting conditions under which each assemblage grew at the expense of the other. Phase relations are shown in Fig. 25. Subsequent work on the $MgSiO_3$-$CaMgSi_2O_6$ join (BOYD and SCHAIRER, 1964), has shown that, for the P-T range investigated in the tremolite study, the pyroxene compositions are approximately $En_{97-98}Di_{03-02}$ and $En_{02-05}Di_{98-95}$. BOYD's 1959 paper was the first to provide experimental P-T diagrams involving synthetic amphibole equilibria. Tremolite is stable up to surprisingly high temperatures, 835° C at 1000 bars fluid pressure, but substitution of Fe^{+2} for Mg lowers the maximum thermal stability as shown by later studies.

GREENWOOD (1962a) has calculated several isobaric T-x curves for devolatilization in the system CaO-MgO-SiO_2-CO_2-H_2O, assuming ideality of mixing in the fluid phase. Some equilibria involving tremolite are shown schematically in Fig. 26. The reactions considered are:

(1) tremolite = 2 diopside + 3 enstatite + quartz + H_2O;
(2) tremolite + 3 calcite + 2 quartz = 5 diopside + $3CO_2$ + H_2O;
(3) 5 dolomite + 7 quartz + H_2O = tremolite + 3 calcite + $7CO_2$; and
(4) serpentine + 9 calcite + $5CO_2$ = tremolite + 7 dolomite + $7H_2O$.

Reaction (1) is a simple dehydration reaction, and is the one experimentally determined by BOYD (1959) under conditions where the fluid phase is virtually pure H_2O; dilution by CO_2 would necessitate a decreased thermal stability range for the hydrous silicate, as discussed previously. Reaction (2) relates a coexisting carbonate + hydrate assemblage to a devolatilized equivalent; because the reaction of tremolite with calcite produces three times as much CO_2 as H_2O, conditions for maximum thermal stability of this assemblage requires equilibration with a fluid phase of approximately 75 mole percent CO_2, assuming ideality of mixing. Reaction (3) relates contrasting hydrate- and carbonate-bearing assemblages to the composition of the fluid phase; increased fugacity of CO_2 at constant total pressure favors the dolomite + quartz compatibility, and increased f_{H_2O} stabilizes tremolite with calcite. Re-

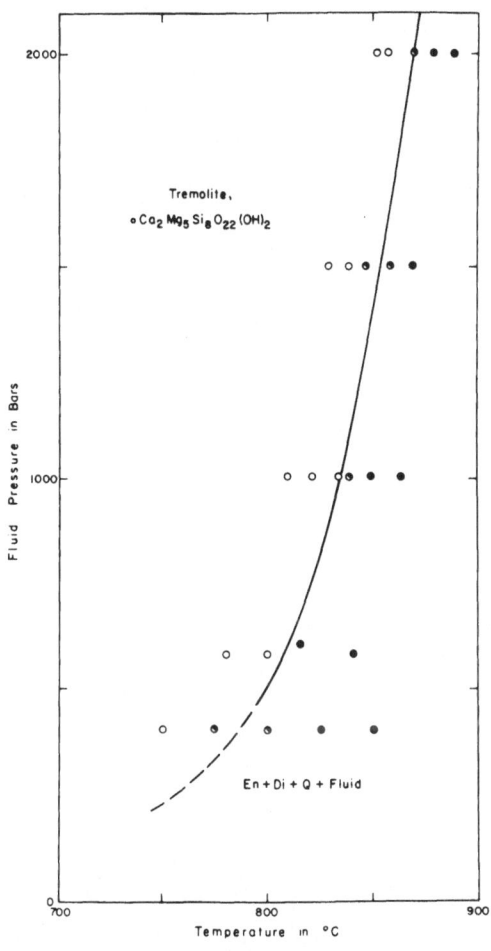

Figure 25

Experimentally determined phase equilibria for tremolite (Boyd, 1959). Divided run symbols indicate persistence of both high and low temperature assemblages of equivalent bulk composition. Abbreviations: En = enstatite; Di = diopside; Q = quartz.

action (4) also involves contrasting hydrate-rich and carbonate-rich assemblages; in this case the tremolite + dolomite compatibility is the CO_2-rich assemblage, serpentine + calcite the H_2O-rich assemblage, so increased f_{CO_2} at constant fluid pressure, *i.e.*, elevated f_{CO_2}/f_{H_2O} ratios, favor the amphibole-bearing assemblage. The T-x relationship of curve

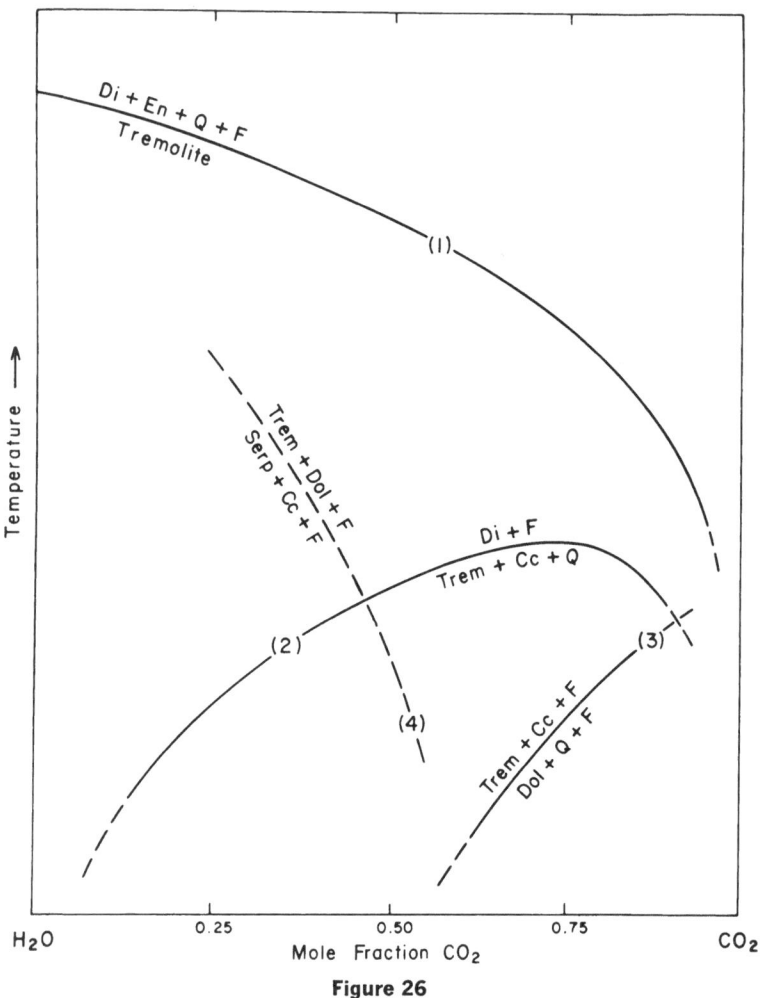

Figure 26

Diagrammatic relationships between isobaric equilibria involving tremolite and the composition of a fluid phase consisting of CO_2 + H_2O (Greenwood, 1962a). Ideality of mixing in the fluid has been assumed. Abbreviations are: Di = diopside; En = enstatite; Q = quartz; F = fluid; Trem = tremolite; Dol = dolomite; Serp = serpentine; Cc = calcite.

(4) respective to both (2) and (3) may not be quite correct; hence this curve is dashed in Fig. 26. Nevertheless, it is topologically satisfactory. Detailed discussions of heterogeneous equilibrium involving multicomponent fluid phase have been presented recently by GREENWOOD (1967) and by EUGSTER and SKIPPEN (1967). A short note has also been presented by METZ (1967).

METZ and WINKLER (1964) have published an experimental isobaric T-x diagram for reaction (2), the conversion of tremolite + calcite + quartz to diopside + fluid, as shown in Fig. 27. The reaction was reversed, demonstrating equilibrium. These authors showed that, at 1000 bars fluid pressure, the thermal maximum for the tremolite + calcite + quartz compatibility lies at 540° C in equilibrium with a fluid phase consisting of approximately 75 mole percent CO_2, in agreement with the calculation of GREENWOOD (1962a). Even at fairly low CO_2 concentra-

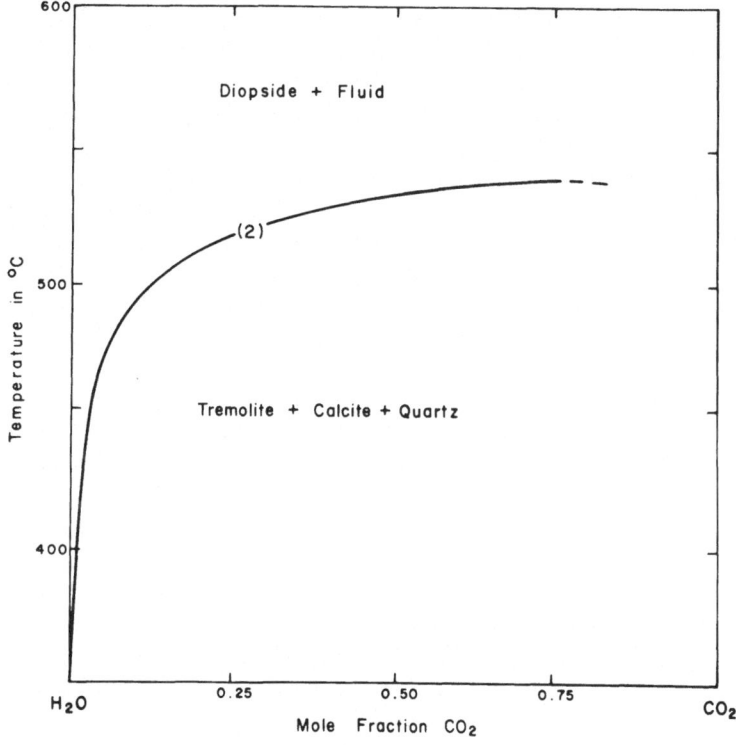

Figure 27

Experimentally determined isobaric equilibria for the reaction tremolite + calcite + quartz = diopside + fluid at 1000 bars (Metz and Winkler, 1964). This curve corresponds to the computed curve (2) of Fig. 26.

tions, say 20 mole percent CO_2 in the fluid phase, diopside is produced from tremolite + calcite only at temperatures in excess of about 500° C.

Eugster *et al.* (1966) have investigated reactions (1) and (2) by controlling fugacities of CO_2 and H_2O. Equilibrium was documented by reversing the reactions. These reactions, and the mineral parageneses of metamorphosed siliceous carbonate rocks have been investigated extensively by Skippen (1967). To a large extent these experimental studies and that by Metz and Winkler have borne out the P-T-x equilibria figured by Greenwood.

Ferrotremolite, $\circ Ca_2 Fe_5^{+2} Si_8 O_{22}(OH)_2$

Stability relations of ferrotremolite were presented by Ernst (1966). Because this mineral contains an element of variable valency, oxygen is involved in some of the reactions encountered in this system. Therefore f_{O_2} is an important variable affecting the phase relations, and must be controlled. The buffer technique of Eugster (1957) was employed (the reader is reminded that buffer curves are shown in Fig. 17). The starting material consisted of an intimate mixture of synthetic ferrotremolite and the high temperature assemblage of equivalent bulk composition appropriate for a specific oxygen fugacity range; equilibrium was demonstrated in the several P-T fields by observing which assemblage grew at the expense of the other. Ferrotremolite is stable only at remarkably low temperatures and under strongly reducing conditions. With oxygen fugacity defined by the magnetite-iron buffer, the thermal stability limit of ferrotremolite lies at 465° C at 1000 bars P_{fluid}, as shown in Fig. 28. The high temperature assemblage of equivalent bulk composition consists of the phases fayalite + quartz + hedenbergitic pyroxene + fluid. At somewhat higher oxidation states with f_{O_2} defined by magnetite + quartz — fayalite buffer (Fig. 29), magnetite joins the high temperature assemblage. For the experimental conditions investigated, the clinopyroxene apparently represents solid solution between $CaFeSi_2O_6$ and $FeSiO_3$ in the range $Hd_{85-95}Fs_{15-05}$. More recent work by Lindsley (1967) indicates that the extent of calcic pyroxene solid solution towards $FeSiO_3$-rich compositions may be less than estimated above.

P-T coordinates for ferrotremolite dehydration curves illustrated in Figs. 28 and 29 are practically identical. The virtual independence of ferrotremolite dehydration and oxygen fugacity in this range is also illustrated in Fig. 30, an isobaric f_{O_2}-T diagram for 3000 bars fluid pressure; this independence arises because the reaction ferrotremolite = fayalite + quartz + hedenbergitic pyroxene + fluid does not involve oxygen.

In contrast, at higher relative oxidation states—say with f_{O_2} values approaching those defined by the bunsenite-nickel buffer—the ferro-tremolite stability field contracts markedly because increased oxygen fugacity strongly favors the more oxidized assemblage of magnetite + quartz + hedenbergitic pyroxene + fluid. At even higher relative oxida-

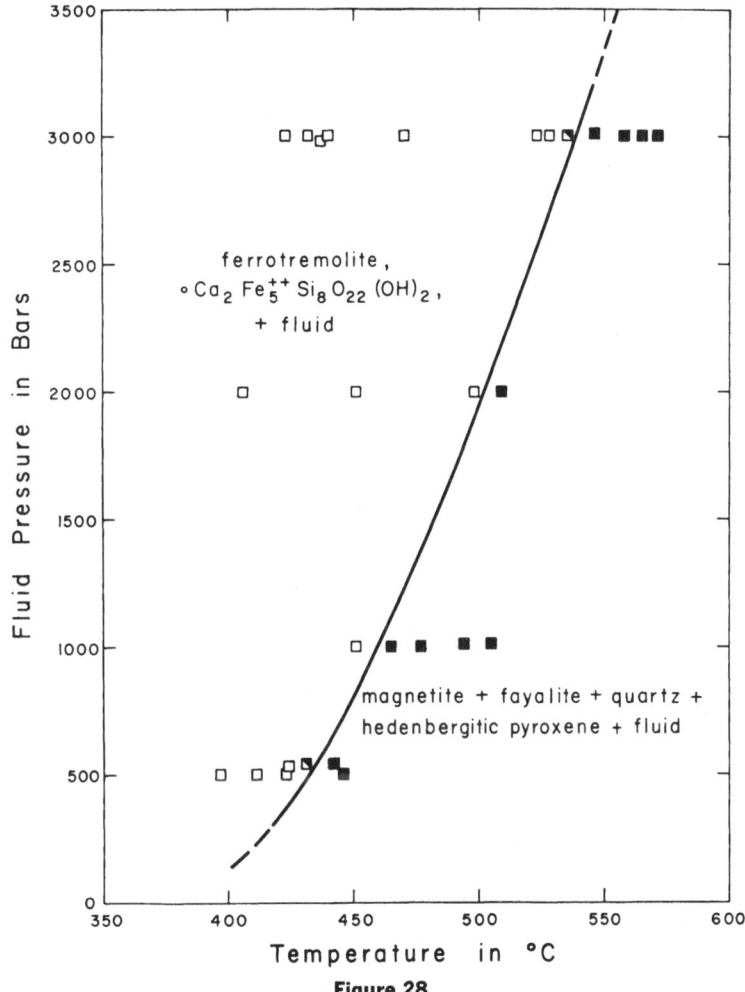

ferrotremolite,
$Ca_2 Fe_5^{++} Si_8 O_{22} (OH)_2$,
+ fluid

magnetite + fayalite + quartz +
hedenbergitic pyroxene + fluid

Figure 28

Experimentally determined stability relations for the ferrotremolite bulk compo-sition, with oxygen fugacities defined by the magnetite-iron and magnetite-wüstite buffers (Ernst, 1966). All paired runs adjacent to the curve constitute reaction reversals. Divided run symbol indicates persistence of both high and low tem-perature assemblages of equivalent bulk composition.

tion states, andradite replaces clinopyroxene in the anhydrous condensed assemblage.

As clearly indicated in Fig. 31, even under favorable oxidation states, the high temperature stability limit of ferrotremolite is 350–370 C° lower than that of tremolite.

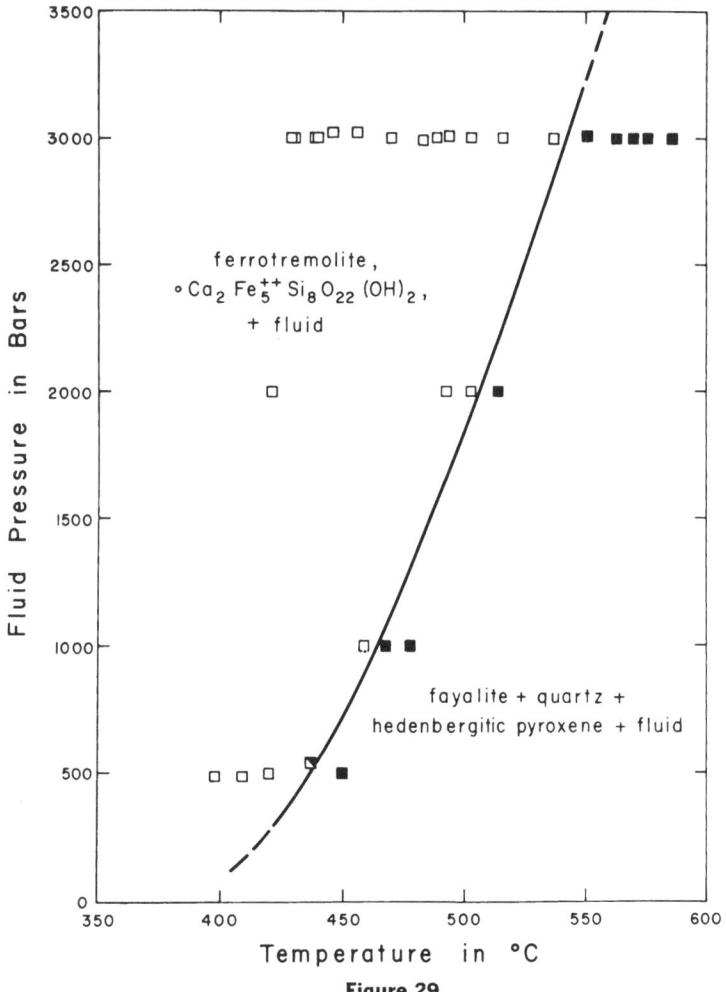

Figure 29

Experimentally determined stability relations for the ferrotremolite bulk composition, with oxygen fugacities defined by the magnetite + quartz − fayalite buffer (Ernst, 1966). All paired runs adjacent the curve constitute reaction reversals. Divided run symbols indicate persistence of both high and low temperature assemblages of equivalent bulk composition.

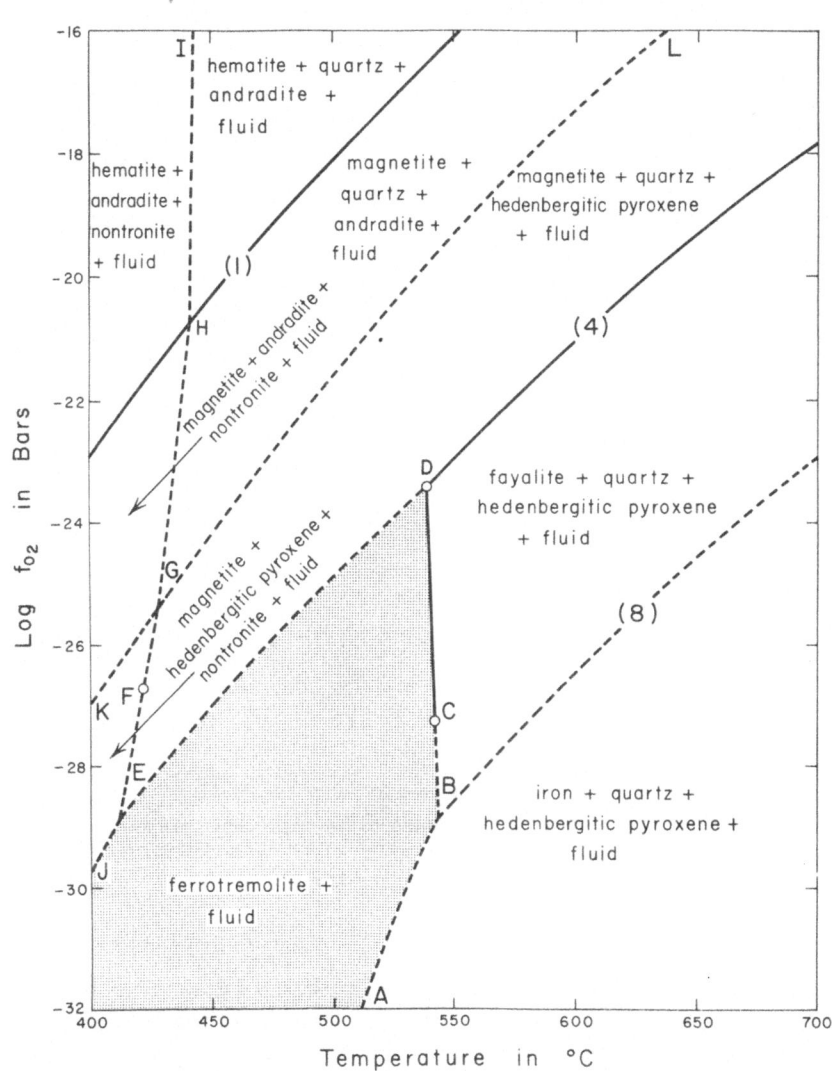

Figure 30

Isobaric log f_{O_2}-T diagram for the ferrotremolite bulk composition at 3000 bars fluid pressure (Ernst, 1966). Field boundaries are dashed where calculated or inferred. Field boundaries (1), (4) and (8) coincide with buffer curves of the same numbers shown in Fig. 17.

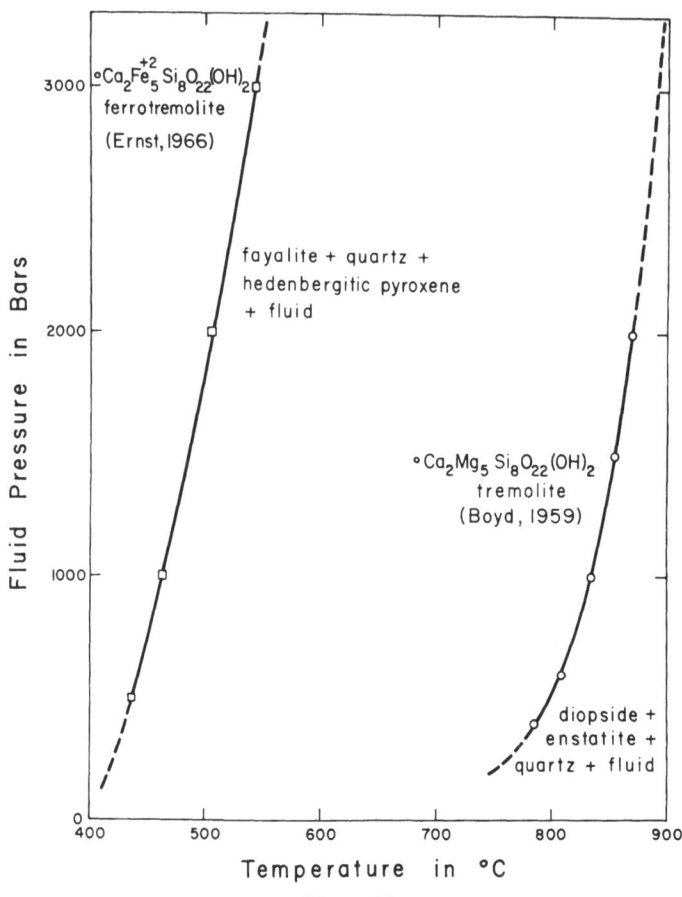

Figure 31

Comparison of the thermal stability limits of tremolite (Boyd, 1959) and ferro-tremolite (Ernst, 1966). For ferrotremolite equilibria, f_{O_2} is defined by the mag-netite + quartz − fayalite buffer.

Actinolites, $\circ Ca_2(Mg,Fe^{+2})_5Si_8O_{22}(OH)_2$

HELLNER and SCHÜRMANN (1966) have investigated the lower thermal stability limit of intermediate members of the $\circ Ca_2(Mg,Fe^{+2})_5Si_8O_{22}$-$(OH)_2$ series, at about 1000 bars P_{H_2O} and 50 bars P_{CO_2}. Experiments were performed using oxalate mixtures + appropriate amounts of water, which maintained a reducing atmosphere. Although they did not control oxygen fugacities, these authors calculated f_{O_2} values slightly less than those defined by the magnetite + quartz − fayalite buffer, and verified their results by running several experiments using a magnetite-iron

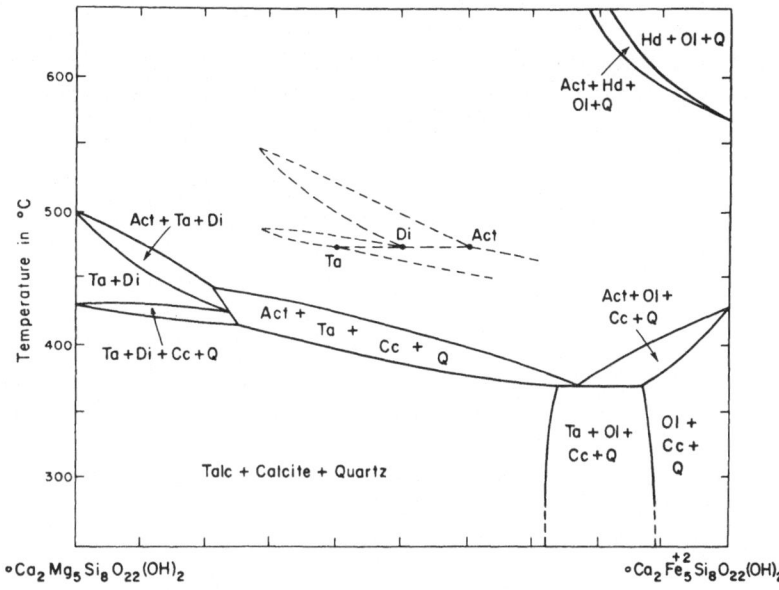

Figure 32

Experimentally determined isobaric T-x diagram for the pseudo-binary join tremolite-ferrotremolite (Hellner and Schürmann, 1966). Values for P_{H_2O} and P_{CO_2} are 1000 and approximately 50 bars respectively. Tentative modification by the author of the phase relations where three transition loops intersect near the composition $Mg_{80}Fe_{20}$ presented by Hellner and Schürmann (1966) is shown as a dashed inset. The abbreviations are as follows: Act = actinolite; Ta = talc; Di = diopside; Cc = calcite; Q = quartz; Ol = olivine; Hd = hedenbergite. All phases coexist with a fluid phase.

buffer.* Temperature-composition relations are shown in Fig. 32. Reactions apparently were reversed. In the presence of CO_2, actinolites decompose at low temperatures to yield calcite + quartz + talc- and/or olivine-bearing assemblages. At intermediate Fe-Mg ratios, both talc and olivine contain considerable amounts of iron. The compatibility of talc + calcite at low temperatures implies that under these experimental conditions, dolomite + quartz cannot coexist. As seen from the reactions 3 dolomite + 4 quartz + H_2O = talc + 3 calcite + $3CO_2$, the dolomite + quartz association apparently is stabilized only at relatively higher ratios of f_{CO_2}/f_{H_2O}. Another point of interest is that the stability range of pure tremolite (and of Mg-rich actinolite) evidently is limited at low

---

*These specific experimental techniques and computations have been discussed recently by MUELLER (1967), who showed the difficulties in estimating gas component fugacities (see also HELLNER and SCHÜRMANN, 1967).

temperatures by the compatibility of talc + diopside according to Fig.
32. The reaction involved here is talc + 2 diopside = tremolite, the
amphibole-bearing assemblage being favored by elevated temperatures;
high pressures, on the other hand, would be expected to restrict the
tremolite stability range in the same manner as that calculated for anthophyllite by GREENWOOD (1963).

At very iron-rich bulk compositions, the high-temperature amphibole
stability limit also has been determined by HELLNER and SCHÜRMANN
(1966). Their value for the conversion of $°Ca_2Fe_5^{+2}Si_8O_{22}(OH)_2$ to
hedenbergitic pyroxene + fayalite + quartz + fluid is about 100 C°
above the value given by ERNST (1966) for corresponding conditions
(1000 bars fluid pressure, oxygen fugacities defined by the magnetite +
quartz − fayalite and magnetite-iron buffers). Provided that CO_2 does
not enter into the amphibole composition, elevated CO_2 pressure at
constant equilibrium pressure of H_2O would cause a slight depression of
the high temperature amphibole stability limit because of the smaller
volume of the anhydrous condensed assemblage: $(\partial P_{total}/\partial T)_{P_{H_2O}} =
\Delta S/\Delta V_{solids}$. Hence the discrepancy between the two works cannot be
resolved as due to 50 bars increment in the total pressure at 1000 bars
P_{H_2O} in experiments by HELLNER and SCHÜRMANN.

A final minor point concerns modification of T-x relationships shown
by HELLNER and SCHÜRMANN for the mutual intersection of the three
transition loops actinolite + talc + diopside, talc + diopside + calcite
+ quartz, and actinolite + talc + calcite + quartz. The three iron-
bearing phases actinolite, talc and diopside have differing Fe/Mg ratios
and can coexist at an isobaric invariant temperature, as illustrated schematically in Fig. 32; presumably actinolite is the most iron-rich of these
phases, talc the most magnesian.

Pargasite, $NaCa_2Mg_4AlSi_6Al_2O_{22}(OH)_2$

High temperature stability relations for pargasite were determined by
BOYD (1959). This is the most refractory hydroxyl-amphibole yet investigated, being stable up to 1040° C at 1000 bars P_{fluid}, and it appears to be
stable under magmatic conditions (see Fig. 33). At fluid pressures less
than about 800 bars, pargasite dehydrates to the assemblage aluminous
diopside + forsterite + nepheline + spinel + anorthite + fluid. Partial
melting of this assemblage at higher temperatures results in the disap-
pearance of nepheline and anorthite. The two melting curves are not
resolved in Fig. 33. At fluid pressures in excess of about 800 bars, pargasite
melts incongruently to the assemblage aluminous diopside + forsterite

+ spinel + liquid + fluid. Addition of excess silica to this system causes production of the condensed assemblage labradorite + diopside + enstatite at high temperatures, and tremolite + plagioclase + minor enstatite at lower temperatures—hence pargasite and free SiO_2 must be incompatible.

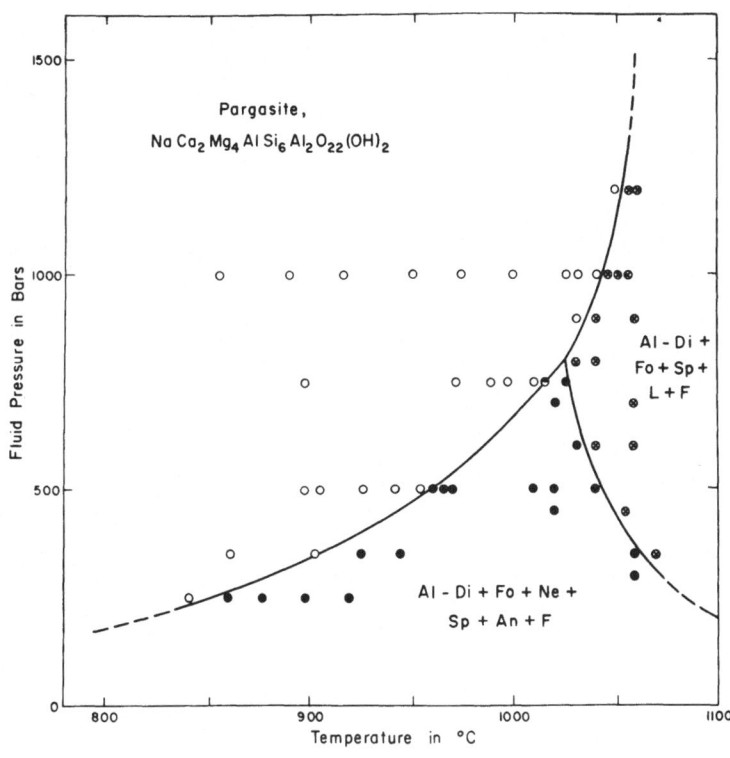

Figure 33

Experimentally determined P_{fluid}-T diagram for the pargasite bulk composition (Boyd, 1959). Divided run symbol indicates persistence of both high and low temperature assemblages of equivalent bulk composition. Abbreviations are as follows: Al-Di = aluminous diopside; Fo = forsterite; Sp = spinel; L = liquid; F = fluid; Ne = nepheline; An = anorthite.

Ferropargasite, $NaCa_2Fe_4^{+2}AlSi_6Al_2O_{22}(OH)_2$

Hydrothermal stability relationships for ferropargasite have been determined by GILBERT (1966) under conditions of controlled oxygen fugacity. Although oxide mixtures were used in many exploratory runs, equilibrium over the entire P-T range investigated was proven by decomposing synthetic amphibole at temperatures slightly in excess of the reaction curve, combined with experiments in which ferropargasite was synthesized from an initial charge consisting of the appropriate high temperature assemblage of equivalent bulk composition. Like ferro-tremolite, ferropargasite has its maximum thermal stability (835° C at 1000 bars fluid pressure) at low oxygen fugacities approaching those

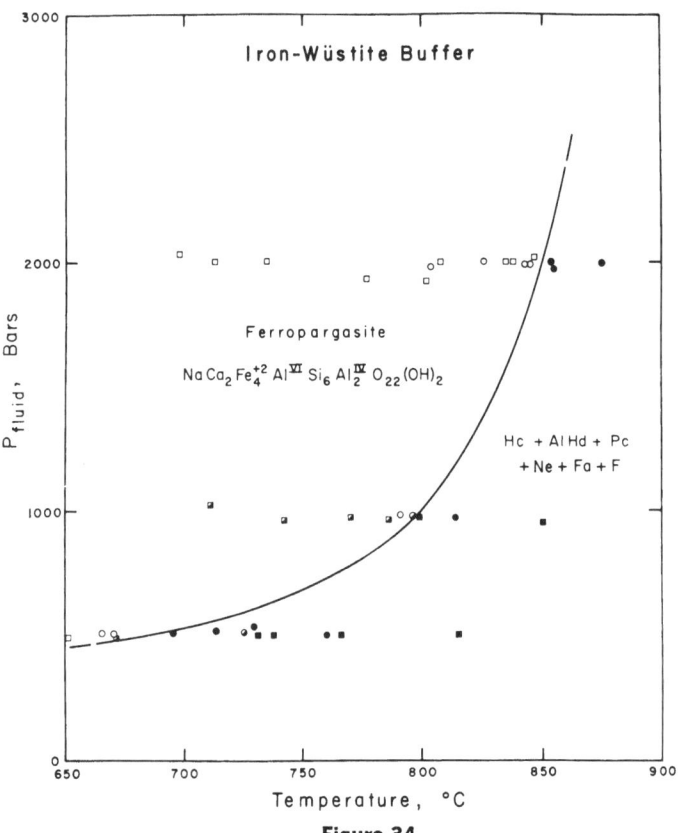

Figure 34

Experimentally determined stability relations for the ferropargasite bulk compo-sition with oxygen fugacities defined by the wüstite-iron buffer (Gilbert, 1966). Square run symbols represent oxide mixture, circles synthetic crystalline starting materials; open symbols indicate a predominantly amphibole-bearing run prod-uct, filled symbols the high temperature assemblage of equivalent bulk compo-sition; divided run symbols indicate persistence of both assemblages; circular run symbols bracketing curves indicate reaction reversals. Abbreviations are: Hc = hercynite-rich spinel; Mt = magnetite-rich spinel; Al Hd = aluminous heden-bergite; Pc = plagioclase; Ne = nepheline; Fa = fayalite; F = fluid; G = garnet.

defined by the wüstite-iron and magnetite-wüstite buffers. Ferropargasite stability relations are shown in Fig. 34, 35 and 36, conventional P_{fluid}-T diagrams at oxygen fugacity ranges specified by the buffers wüstite-iron, magnetite + quartz − fayalite, and bunsenite-nickel respectively. The ferropargasite thermal stability limit is depressed more than 300 C° with

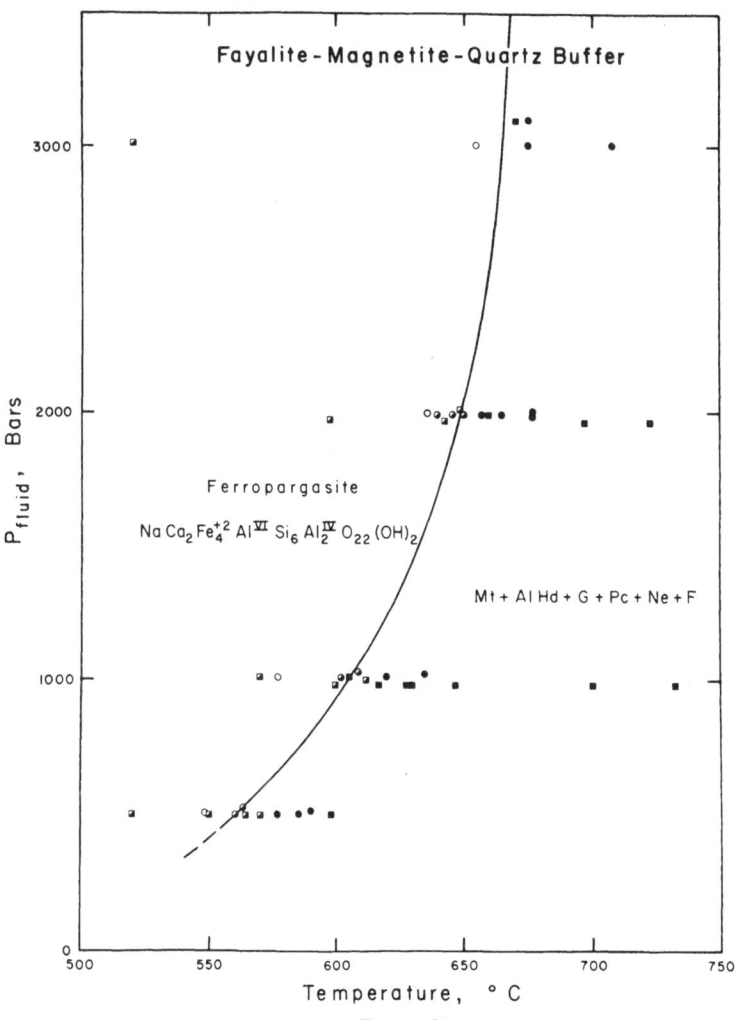

Figure 35

Experimentally determined stability relations for the ferropargasite bulk composition with oxygen fugacities defined by the magnetite + quartz − fayalite buffer (Gilbert, 1966). Run symbols and abbreviations are identical to those of Fig. 34.

increasing oxidation state, and the reduced high temperature assemblage of hercynite + aluminous hedenbergite + plagioclase + nepheline + fayalite + fluid is replaced by a more oxygen-rich compatibility, magnetite + aluminous hedenbergite + garnet + plagioclase + fluid.

The effects of variation in oxygen fugacity on the investigated phase equilibria are illustrated in the isobaric f_{O_2}-T diagram at 2000 bars fluid

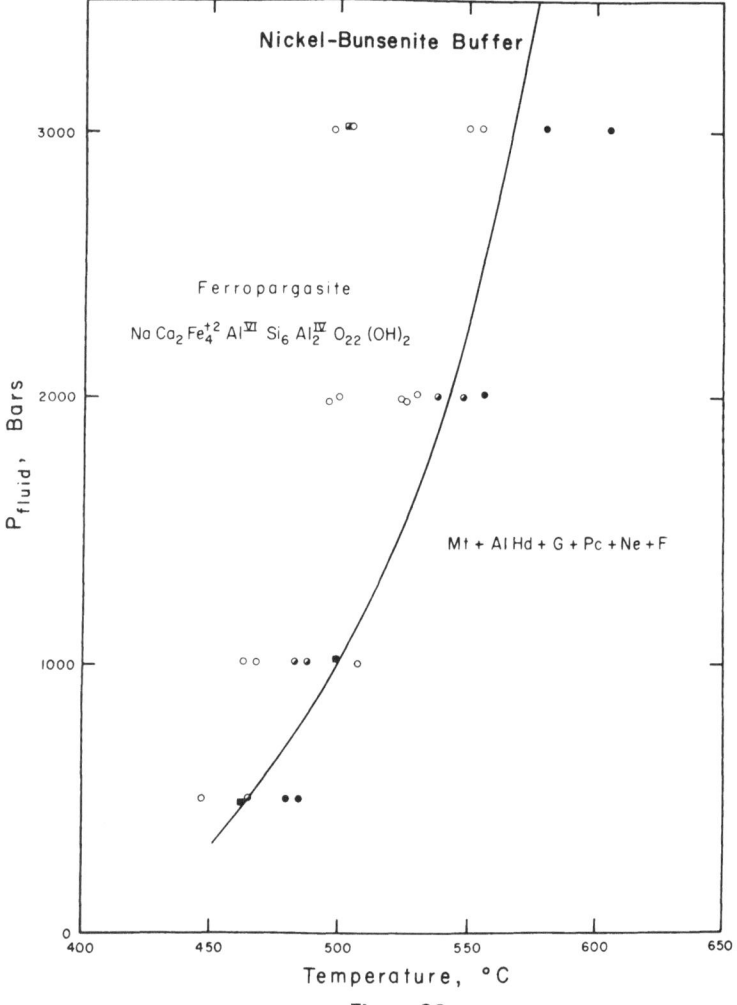

Figure 36

Experimentally determined stability relations for the ferropargasite bulk composition and oxygen fugacities defined by the bunsenite-nickel buffer (Gilbert, 1966). Run symbols and abbreviations are identical to those of Fig. 34.

pressure (Fig. 37). The curve separating garnet-bearing and fayalite-bearing high temperature assemblages equivalent in bulk composition to ferropargasite intersects the amphibole dehydration curve at its isobaric thermal maximum, point D in Fig. 37; the spinel phase labeled Hc is hercynite-rich spinel in the more reduced high temperature assemblage, Mt is magnetite-rich spinel in the more oxidized phase compatibility, but GILBERT (1966, p. 732–733) noted that in this f_{O_2}-T region it is a homogeneous hypersolvus phase (see TURNOCK and EUGSTER, 1962). It is clear from the diagram that at elevated oxidation states, garnet + magnetite-bearing associations are strongly favored over ferropargasite. Moreover, at very low values of f_{O_2}, the hydrogen content of the fluid phase becomes so large that f_{H_2O} is severely curtailed, accounting for the fact that the hydrous condensed assemblage, amphibole, contracts at extremely low oxidation states.

As evident from Fig. 38, the thermal stability limit of ferropargasite is reduced more than 200 C° compared with that of pargasite, even with oxygen fugacities most favorable for stabilization of the iron end-member.

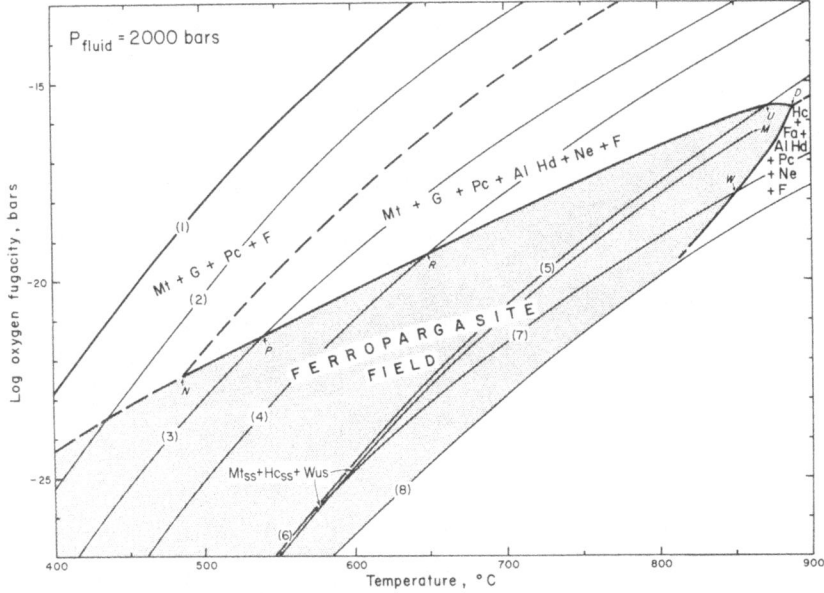

Figure 37

Isobaric log f_{O_2}-T diagram for the ferrotremolite bulk composition at 2000 bars fluid pressure (Gilbert, 1966). Field boundaries are dashed where calculated or inferred. The numbered buffer curves correspond to those of Fig. 17. Abbreviations are identical to those of Fig. 34.

Reconnaissance experiments involving the addition of excess silica to the ferropargasite bulk composition showed that this compositional change apparently causes the high temperature amphibole stability limit to decline at low relative oxidation states compared to that for pure ferropargasite, but to be extended at higher oxidation states; GILBERT

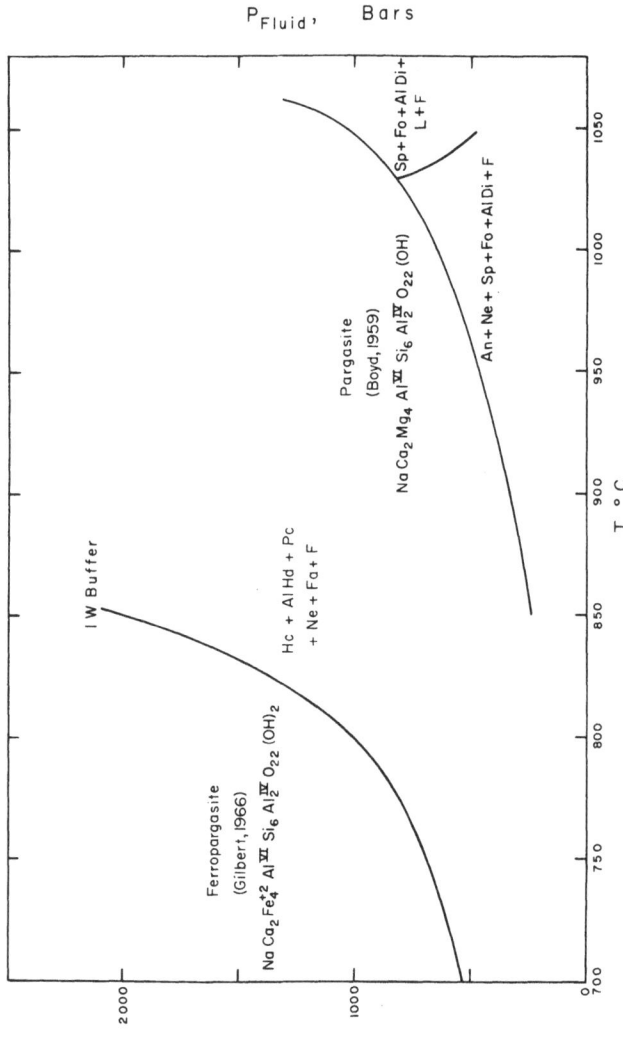

Figure 38

Comparison of the thermal stability limits of pargasite (Boyd, 1959) and ferropargasite (Gilbert, 1966). For ferropargasite equilibria, f_{O_2} is defined by the wüstite-iron buffer. Abbreviations are the same as those of Figs. 33 and 34.

suggested that, under the latter conditions, the amphibole had become more siliceous.

Other Calcic Amphiboles

Although phase equilibrium relations have not been demonstrated for other calcic amphiboles, syntheses by Colville et al. (1966) suggest the feasibility of further experimental study. The physical conditions of synthesis *might* reflect the existence of stability fields. Run conditions for the several amphiboles were:

edenite	$850°$ C, 2000 bars P_{fluid}, no buffer;	
ferroedenite	$600°$ C, 3000 bars P_{fluid}, wüstite-iron buffer;	
magnesiohastingsite	$850°$ C, 2000 bars P_{fluid}, hematite-magnetite buffer;	
and hastingsite	$600°$ C, 3000 bars P_{fluid}, magnetite + quartz — fayalite buffer.	

Efforts to synthesize tschermakite and ferrotschermakite have been unsuccessful thus far.

The fluorine- and boron-bearing calcic amphiboles fluortremolite, fluoredenite, and boron edenite have been synthesized by Comeforo and coworkers (Comeforo and Kohn, 1954; Kohn and Comeforo, 1955; Shell et al., 1958) but phase relations were not determined. Starting materials employed were mixtures of oxides and fluorides; amphiboles were grown from slowly cooled melts at one atmosphere total pressure.

Schürmann (1966, p. 28—35) synthesized aluminous actinolites using oxalate-bearing oxide mixtures as starting materials; compositions included $Ca_2(Mg,Fe^{+2})_{4.5}Al_{0.5}Si_{7.5}Al_{0.5}O_{22}(OH)_2$ and iron-rich members of the series $Ca_2(Mg,Fe^{+2})_4AlSi_7AlO_{22}(OH)_2$.

Natural Occurrences of the Calcic Amphiboles

Among the amphiboles, those containing Ca in the M_4 structural site are the most abundant, both from the point of view of distinguishable species and absolute amounts. They occur in a wide variety of geologic environments, including marbles, low and medium grade regional and contact metamorphic types, as a primary constituent of plutonic igneous rocks, and less commonly in some volcanics. Because these amphiboles are so widespread and diverse in their occurrence, only a superficial treatment can be presented here.

Tremolite-actinolite is a characteristic phase encountered in the decarbonation of siliceous dolomite. It probably is generated through the reactions:

5 dolomite + 8 quartz + H_2O = tremolite + 3 calcite + $7CO_2$;
and 4 serpentine + 9 calcite + $5CO_2$ = tremolite + 7 dolomite + $7H_2O$,

as illustrated in Fig. 26, reactions (3) and (4). At very low CO_2 fugacities, the reaction:

6 calcite + 5 talc + 4 quartz = 3 tremolite + $2H_2O$ + $6CO_2$

could also produce tremolite, as discussed by RAMBERG (1944) and experimentally synthesized by HELLNER and SCHÜRMANN (1966). These are among the lowest grade reactions recognized in marbles. BOWEN (1940, p. 260) listed tremolite as the first of ten index minerals whose order of appearance for appropriate bulk compositions represents the onset of higher temperature and/or lower CO_2 pressure conditions. At higher grades, tremolite reacts with other phases to form successively less volatile-rich assemblages:

11 dolomite + tremolite = 8 forsterite + 13 calcite + $9CO_2$ + H_2O;
3 calcite + 2 quartz + tremolite = 5 diopside + $3CO_2$ + H_2O;
5 calcite + 3 tremolite = 11 diopside + 2 forsterite + $5CO_2$ + H_2O;
tremolite + forsterite = 2 diopside + 5 enstatite + H_2O;

and finally, tremolite = 2 diopside + 3 enstatite + quartz + H_2O. This last reaction is the one experimentally investigated by BOYD (1959). With the exception of the two-pyroxene (+ quartz) assemblage, all significant compatibilities are shown by BOWEN (1940, Figs. 2–4). Epidote + plagioclase + aluminous actinolite skarns have been described by FRANCIS (1958). Edenite and pargasite have also been reported from calc-silicate parageneses.

As indicated by experimental study (ERNST, 1966), the extreme rarity of ferrotremolite reflects the low thermal stability range of this mineral, its restriction to relatively reducing environments, and the departure of rock bulk compositions from that of the mineral itself. A low fugacity of H_2O would also disfavor this mineral. Amphiboles most closely approaching the ferrotremolite composition occur in feebly metamorphosed iron formations (MUELLER, 1960).

Although actinolites and blue-green hornblende have been described from glaucophane schists of Japan (SUZUKI, 1930; IWASAKI, 1963; BANNO, 1964) and California (SWITZER, 1951; BORG, 1956; COLEMAN and LEE, 1963; ERNST, 1965), the calcic amphiboles are most abundant in the more "normal" metamorphic sequences. Actinolites are nearly ubiquitous in greenschist facies mafic metaigneous rocks. The typical assemblage consists of actinolite + epidote or pumpellyite + albite + chlorite + sphene (± calcite) as documented in Scotland by WISEMAN (1934, p. 368) and TILLEY (1938), New Zealand (TURNER, 1935; HUTTON, 1940) and Japan (MIYASHIRO, 1958, p. 247; BANNO, 1964), among others. These actinolites characteristically contain 2–4 weight percent Al_2O_3, and about a percent of Na_2O. With increasing metamorphic grade, these constituents increase in certain cases, principally at the expense of silica content of the amphibole (HARRY, 1950; ENGEL and ENGEL, 1962; BINNS, 1965b; LEAKE, 1965).

In some terranes, for instance the Scottish Highlands, the progressive metamorphic sequence is marked by a regular, continuous change in amphibole composition from actinolite → blue-green hornblende → olive-brown hornblende in rocks of basaltic composition (WISEMAN, 1934). However, SHIDO and MIYASHIRO (1959, p. 94) described two Scottish Highland epidiorites which contain parallel and irregular intergrowths of actinolite and hornblende; in these cases the grain boundaries are sharp but it is not clear whether or not the two amphiboles are in mutual equilibrium. Other terranes such as the Abukuma Plateau, Japan (SHIDO, 1958; SHIDO and MIYASHIRO, 1959) exhibit the metabasalt paragenesis: actinolite → actinolite + hornblende → olive-brown hornblende. Such two Ca-amphibole assemblages imply the existence of a solvus, provided the phases are in mutual equilibrium. As discussed by MIYASHIRO (1961), the Abukuma Plateau was affected by a relatively low pressure, high temperature metamorphic event (andalusite-sillimanite type) compared to the Scottish Highlands recrystallization (kyanite-sillimanite type). Thus it is conceivable that the crest of the "hornblende solvus" was depressed relative to the compositional migration of amphibole in response to P-T gradients attending Barrovian progressive metamorphism. Apparently in the contrasting Abukuma terrane, at the entrance to amphibolite facies conditions, the migration of amphibole composition in response to increasing metamorphic grade intersected the two-phase field.

Typical parageneses from the Abukuma Plateau and Scotland are reproduced in Figs. 24 and 39 respectively. Apart from the differences in calcic amphibole parageneses, the two metamorphic sequences differ in that mafic rocks of the kyanite-sillimanite type preserve sodic plagioclase and epidote to higher grade, and contain almandine garnet in all but the most feebly recrystallized rocks; in comparison, in the andalusite-sillimanite type, cummingtonite, pyroxenes and calcic plagioclase appear at middle grades and, with the exception of the Fe-Mg amphibole, increase in the higher grade rocks.

Although compositions of amphiboles in progressive metamorphic sequences generally trend from actinolite towards hastingsite, other substitutions reflect departure of rock bulk compositions from that of the mafic igneous rocks. Hence somewhat tschermakitic amphiboles occur in calcareous metapelites of Britain and the Alps (TILLEY, 1937, Table IV), and edenite-pargasites have been reported from anorthositic metanorites in India (SUBRAMANIAM, 1956). In general, rock bulk composition appears to exert a more pronounced control over the chemistry of the amphibole than do the physical conditions. It is true however that with increased temperatures, the maximum amount of Ti incor-

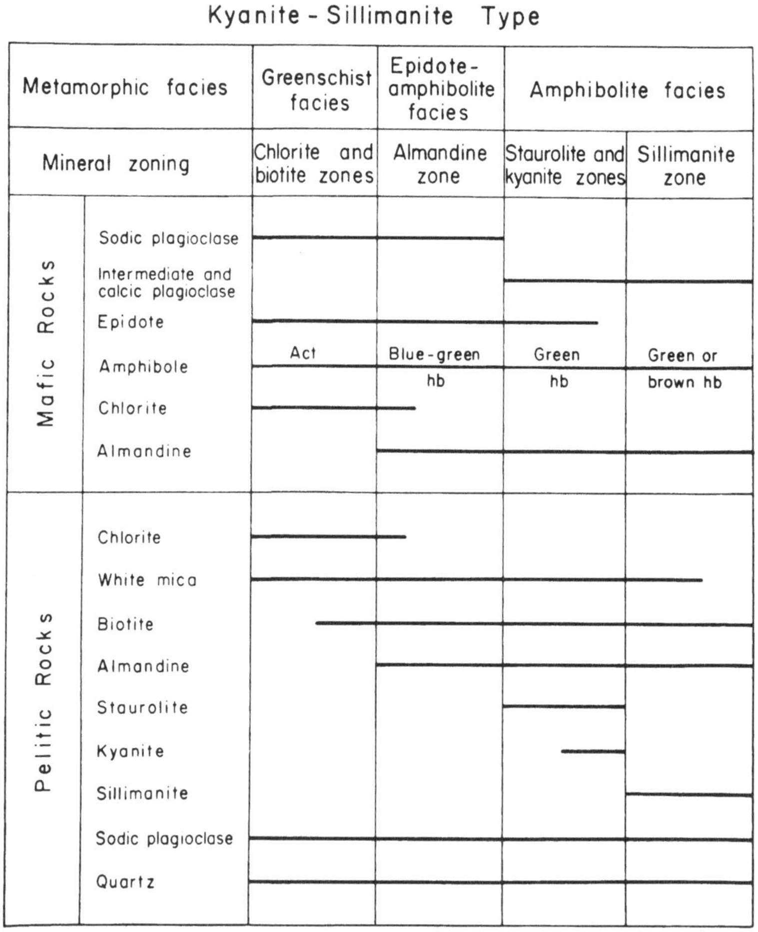

Kyanite - Sillimanite   Type

Metamorphic facies		Greenschist facies	Epidote-amphibolite facies	Amphibolite facies	
Mineral zoning		Chlorite and biotite zones	Almandine zone	Staurolite and kyanite zones	Sillimanite zone
Mafic Rocks	Sodic plagioclase				
	Intermediate and calcic plagioclase				
	Epidote				
	Amphibole	Act	Blue-green hb	Green hb	Green or brown hb
	Chlorite				
	Almandine				
Pelitic Rocks	Chlorite				
	White mica				
	Biotite				
	Almandine				
	Staurolite				
	Kyanite				
	Sillimanite				
	Sodic plagioclase				
	Quartz				

Figure 39

Mineral parageneses in progressive metamorphism of the Scottish Highlands (Wiseman, 1934; Miyashiro, 1961).

porated is greater than at low temperatures, and that at elevated pressures the maximum amount of Al^{VI} in amphiboles increases (LEAKE, 1965).

Oxygen fugacity is an important variable which must be considered in the paragenesis of calcic amphiboles. At constant temperature and fluid pressure, garnet-rich rocks would appear to be favored at relatively high oxidation states, hornblende-rich rocks at lower oxidation states, according to the recent studies by ERNST (1966), and GILBERT (1966); such relations are evident from Figs. 30 and 37. The composition of the garnet, too, is affected by the oxygen fugacity. Although high values of f_{O_2} favor andradite, low oxidation states are required for the stabilization of almandine (HSU, in press).

Metamorphic aureoles surrounding epi- and mesozonal plutons display amphibole parageneses in mafic rocks somewhat similar to those of regional metamorphism (e.g., see BINNS, 1965a). COMPTON (1958) described the rather abrupt conversion of typical actinolite-bearing greenschist country rocks to a hornblende + andesine $\pm$ clinopyroxene hornfels adjacent to the contact with a small batholith in California. COMPTON emphasized the rate of heating and role of reaction kinetics to account for overstepping physical conditions appropriate to the albite-epidote amphibolite facies. Alternatively, relatively high temperatures and low pressures might result in a calcic amphibole miscibility gap; such parageneses may therefore be more comparable to those of the Abukuma (andalusite-sillimanite type) than to those of the Barrovian style of metamorphism.

In igneous rocks, calcic amphiboles occur most abundantly in intermediate and felsic varieties, although mafic rocks have bulk compositions closer to those of the various amphiboles than do the felsic rocks. This is because subsilic magmas are typically undersaturated with respect to H_2O (and F_2), hence the fugacities of volatile components are too low to stabilize OH- and F-bearing minerals at magmatic temperatures. Where hornblendes occur as primary phases in basic rocks they are Mg-rich, thus are relatively refractory; amphiboles of ultramafic rocks are typically pargasitic hornblendes (e.g., see ONUKI, 1964). The intermediate and silicic rocks have lower solidus temperatures and contain much greater dissolved volatiles, so hornblende can coexist with melt in a great proportion of these rocks.

Hastingsitic amphiboles are most abundant in calc-alkaline diorites, granodiorites and granites (BILLINGS, 1928a, b; BUDDINGTON and LEONARD 1953; ENGEL, 1959; LIPMAN, 1964). The occurrence of arfvedsonite in some such plutons, hastingsite in others has been related to the rock bulk composition (ERNST, 1962, p. 733); higher An/Ab ratios favor the calcic amphibole as shown in Fig. 40. (See also BORLEY, 1963; FROST, 1963; BORLEY and FROST, 1963.)

Although phase relationships have not been determined experimentally for magnesiohastingsite and hastingsite, P-T conditions for their stabilities probably are similar to those of the pargasite-ferropargasite series, by analogy with P-T relations determined for sodic amphibole Fe^{+3} and Al^{VI} end-members. However, because the hastingsites contain ferric as well as ferrous iron, iron-rich members of this group undoubtedly have a more extensive f_{O_2}-T stability range than ferropargasitic amphibole. Because it is also probable that the more hydrous intermediate and silicic igneous rocks maintain relatively high oxygen fugacities due to assimilation of crustal material containing more nearly stoichiometric H_2O, primary hastingsitic amphiboles are strongly favored over pargasitic varieties.

From the hydrothermal investigations by YODER and TILLEY (1962, p. 440–470), it is apparent that amphibolites are stable to high temperatures under conditions where fluid and lithostatic pressures are equal; in

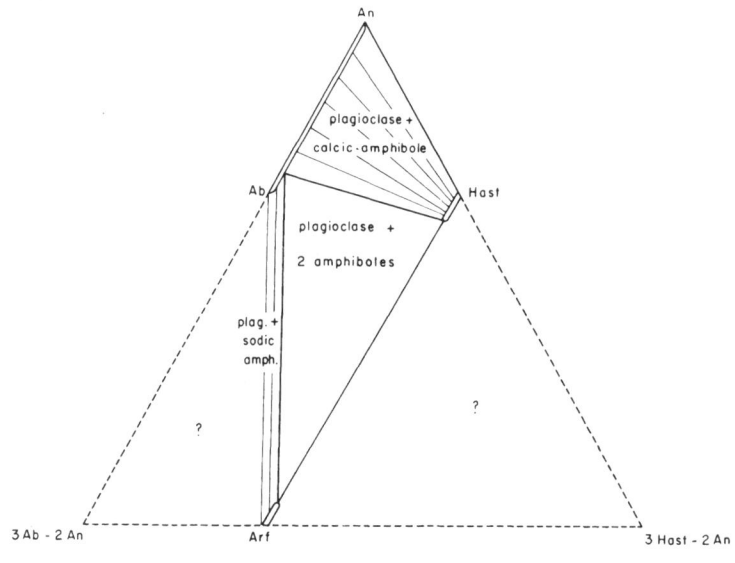

mole   percent

Figure 40

Schematic isothermal, isobaric phase relations in the quaternary system H_2O-$NaAlSi_3O_8$(Ab) — $CaAl_2Si_2O_8$(An) — $NaCa_2Fe_4^{+2}Fe^{+3}Si_6Al_2O_{22}(OH)_2$(Hast) extended to negative values for An, and projected onto the plane An-(3Ab − 2An)-(3Hast − 2An) (Ernst, 1962). The composition of arfvedsonite, $Na_3Fe_4^{+2}Fe^{+3}Si_8$-$O_{22}(OH)_2$ may be represented in this system as 2Ab+Hast−2An. Limited amphibole solid solution is assumed, and all phases are considered as stable with fluid.

fact, depending on the bulk composition of the rock, hornblende is in equilibrium with melt at fluid pressures above 1000–2000 bars and temperatures approaching 900–1000° C (see Fig. 41). The aluminous magnesian calcic amphiboles are, of course, stable at higher temperatures than their Fe^{+2} analogues (compare the experimental curves of Boyd, 1959, and Gilbert, 1966, as shown in Fig. 38). In deeper levels of the

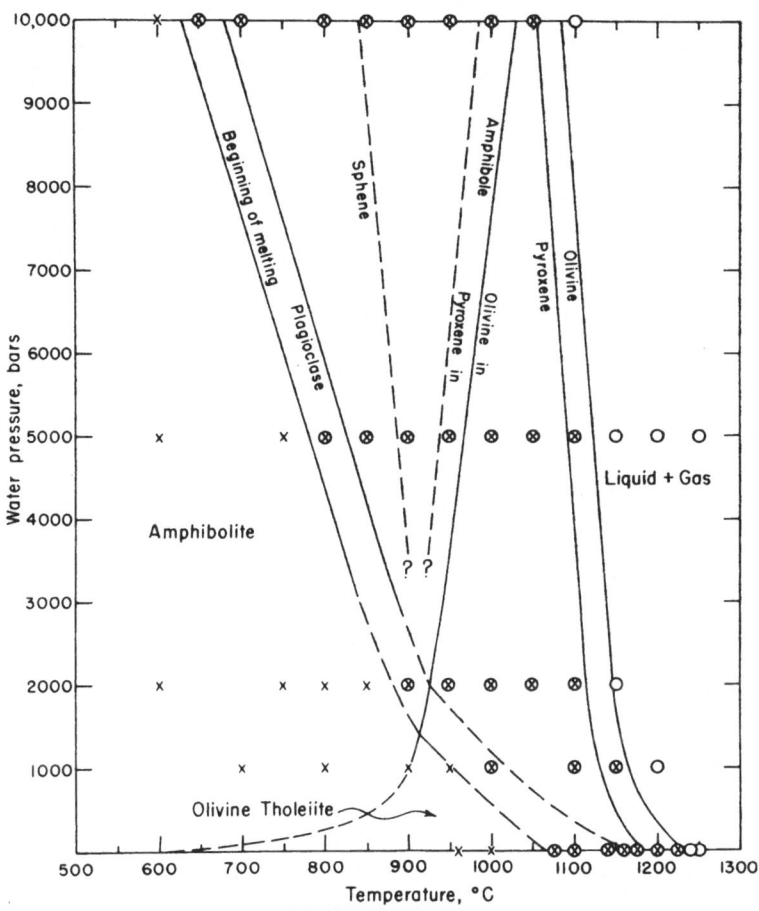

Figure 41

Experimentally determined phase relationships for a sample of natural olivine tholeite (Yoder and Tilley, 1962, Fig. 27). The thermal stability range of amphibole is not markedly affected starting with high-alumina basalt, alkali basalt or oxidized hawaiite. Run symbols are as follows: open circles = liquid; crossed circles = liquid + crystals; crosses = crystals. All condensed phases coexist with fluid.

earth's crust, provided aqueous fluid is abundant, interlayered amphibolites and feldspathic migmatities are to be expected to result from the partial melting of mafic, pelitic and felsic preexisting rocks (see also TUTTLE and BOWEN, 1958, p. 122–125; WINKLER, 1965, p. 176–208).

YAMAZAKI et al. (1966) have described tschermakitic hornblendes from gabbro inclusions in calc-alkaline volcanics of central and northern Honshu, Japan. These authors drew attention to the compositional similarity of these inclusions to kyanite eclogites. Based on the anorthite + tschermakite ± Al-pyroxene mineral assemblage, and the more aluminous nature of the amphibole compared to those of subsilicic plutonic rocks, YAMAZAKI et al. suggested that these inclusions were derived from the vicinity of the Mohorovicic Discontinuity.

MASON (1966) described somewhat similar Ti+Al-rich alkaline hornblende xenocrysts which occur in a New Zealand volcanic breccia associated with xenocrysts of pyropic garnet and Al-augite; in addition to these phases, the breccia contains nodules of eclogite and dunite. Judging from the association, MASON suggested that the xenocrysts originated deep within the crust or in the upper mantle.

Amphiboles in the Mantle

GREEN and RINGWOOD (1963) discussed the mineralogy of the mantle, basing their model on apparent equilibrium assemblages observed as inclusions in kimberlites, intrusive peridotites and basalts. They suggested that the following assemblages reflect increasing depths below the Mohorovicic Discontinuity:

(1) olivine + amphibole + minor Cr-spinel;
(2) olivine + plagioclase + enstatite + clinopyroxene + minor Cr-spinel;
(3) olivine + Al-enstatite + Al-clinopyroxene + spinel;
and (4) olivine + pyrope + pyroxene(s).

GREEN and RINGWOOD assumed a mantle composition, termed pyrolite, consiting of three parts dunite and one part basalt; they pointed out that under the conditions of low fluid pressure with temperatures approaching 500° C thought to exist in the uppermost suboceanic mantle, the amphibole-bearing assemblage should be stable.

Based on the disparity between relatively high K_2O contents of basalts presumably derived from mantle ultramafics, and the very low levels of potassium incorporated in anhydrous mantle phases such as olivine, orthopyroxene, clinopyroxene, spinel and garnet, OXBURGH (1964) also postulated the presence of a K-bearing minor constituent, making up 2–20 weight percent of the parent upper mantle. He further argued that this phase was hornblende, based on experimentally determined incongruent melting relations of layer-lattice and chain silicates, as well as on the ratios of K_2O/Na_2O of the various basaltic magma types. In this scheme, the low velocity layer would represent undifferentiated amphibole-bearing ultramafic; the uppermost "barren" mantle would be dunite (correlated with olivine nodules included in alkali basalts) which was left as a crystalline residue subsequent to fractional melting.

HART and ALDRICH (1967) have shown that potassium is strongly concentrated in amphiboles with respect to rubidium. According to these authors, the high K/Rb ratios of abyssal tholeiites might reflect incongruent melting of amphibole peridotite as postulated by OXBURGH (1964); in any case, all arguments regarding mantle compositions which are based on trace (and major) element abundances ought to take into account the control that mineral equilibria exert on fractionation.

MELSON et al. (1967) have described pargasite + olivine + enstatite + Cr-spinel as primary constituents of ultramafic mylonites cropping out on St. Peter and St. Paul Rocks. The amphibole occurs both as augen and as a groundmass mineral; textural relations and the intimate interlayering of these rocks with diopside-rich mylonites indicate that at least some of the pargasites are not secondary. MELSON et al. consider these mylonites to have originated in the upper mantle; because of their remoteness from continental crust, such rocks are unlikely to have undergone significant contamination during emplacement. According to CLARK and RINGWOOD (1964) amphibole-bearing assemblages of this nature should be stable in the upper 30 km of the suboceanic mantle, assuming a relatively low activity of H_2O. However, as pointed out later by RINGWOOD (1966), this hydrous assemblage could persist to greater depths provided high activities of H_2O were maintained locally.

The only high pressure experimental investigation bearing on the P-T stability of amphiboles at high pressures is that involving solic amphibole polymorphism (ERNST, 1963a). Apparently these amphiboles are stable at least up to pressure of 30–40 kb. GREENWOOD (1963) calculated that at 600–800° C anthophyllite should have a high pressure stability limit of about 20–25 kb, as illustrated in Fig. 21. Unfortunately, there are no high pressure data for the calcic amphiboles. Judging from the crystal structures, one might predict that at very high pressures amphiboles would transform to hydrous single-chain type structures, provided other heterogeneous reactions do not intervene. The basis for the hypothesized transition is that amphiboles have more open structures, hence lower densities than do the pyroxenes, as indicated by the sizes of sites and cation coordinations:

$$Mg_8^{VI}Si_8^{IV}O_{24} \text{ (s.g.} = 3.21) \text{ vs. } o_1^X Mg_7^{VI}Si_8^{IV}(O+OH)_{24} \text{ (s.g.} = 2.85);$$
$$\text{and } Ca_4^{VIII}Mg_4^{VI}Si_8^{IV}O_{24} \text{ (s.g.} = 3.22) \text{ vs. } o_1^X Ca_2^{VIII}Mg_5^{VI}Si_8^{IV}(O+OH)_{24}$$
$$\text{(s.g.} = 3.02).$$

However, the presence of hydrogen in the amphibole reflects the electrostatic environment of the O_3 anionic site, itself a function of the double-chain geometry. In the double-chain silicates, the occupant of $O_3(OH^-)$ is bonded to octahedrally coordinated cations, and has no bond with Si. In contrast, among the single-chain silicates, all anions are bonded to at least one tetrahedrally coordinated cation, hence the requirement of local electrostatic charge balance might prohibit the anionic acceptance of protons. Another aspect of this relationship is that, among the amphiboles, the O-H vector $a^*(=a \sin \beta)$ is normal to the plane of the six-fold coordinated cations, and is directed toward the A structural site (BURNS and STRENS, 1966); thus the proximity of the proton, presumably repulsed by the VI-coordinated cations, undoubtedly influences the electronic configuration of the occupancy of the A site (GIBBS and PREWITT, 1966). No analogous position exists among single-chain structures, so the existence of OH^- in pyroxene-like minerals is entirely speculative at this stage.

In this regard it is interesting to note that SCLAR et al. (1967) reported the synthesis of Mg-pyroxene containing 20.4 weight percent H_2O at pressures between 75–180 kb in the temperature range 525–750° C. The structural formula of this phase was deduced to be $Mg_2Si(H_4O_4)O_2$, where four protons have replaced one silicon, as in hydrogrossular.

Chapter VII.

EXPERIMENTAL PHASE RELATIONS AND OCCURRENCE OF THE SODIC AMPHIBOLES

Glaucophane, $\square Na_2Mg_3Al_2Si_8O_{22}(OH)_2$

Hydrothermal phase equilibria for glaucophane and for glaucophane + quartz were determined by ERNST (1961). Starting materials were oxide mixtures in the appropriate stoichiometric proportions or glasses ($+H_2O$) in most runs, but reaction reversals were demonstrated by recrystallizing synthetic crystalline assemblages, both at low and high fluid pressures. Charges of the bulk compositions $Na_2O \cdot 3MgO \cdot Al_2O_3 \cdot 8SiO_2 + H_2O$ and $Na_2O \cdot 3MgO \cdot Al_2O_3 \cdot 10SiO_2 + H_2O$ never were crystallized completely to amphibole, but judging from the proportions among the anhydrous phases utilizing both x-ray and micrometric techniques, the hydrous phase was estimated to be essentially pure glaucophane. This synthetic amphibole has a unit cell volume roughly two percent greater than that of natural glaucophane. The experimentally determined ΔH for breakdown of this amphibole is approximately $325 + 60$ kcal/mole, almost an order of magnitude larger than for dehydration reactions involving other OH-bearing silicates; such anolamous values for enthalpy change arise as a consequence of the steep dP/dT slope for the glaucophane thermal stability limit. The high temperature stability limit of this mineral is surprisingly high, $864 \pm 5°$ C at 1000 bars fluid pressure, judging from its natural restriction to relatively low temperature metamorphic rocks. All these observations call into question the validity of relations presented in Figs. 42 and 43; nevertheless, equilibrium was demonstrated for the glaucophane bulk composition at 820 and 2000 bars and for the glaucophane + quartz bulk composition at 1500 bars by reversing the reactions.

Complex phase relationships at high temperatures reflect reaction among crystals, fluid and melt; because such interactions exert little control over amphibole equilibria in nature, they are not treated here and the interested reader is referred to the original paper.

With the addition of excess silica, the thermal stability limit of glaucophane is lowered only 3–6 C°. This phenomenon results in part from the fact that the reaction forsterite + quartz = 2 enstatite involves only a small percentage decrease in Gibbs free energy of the high temperature assemblage relative to glaucophane; moreover synthetic glaucophane

77

apparently has an unusually low entropy. Therefore, the intersection of G-T curves for glaucophane + quartz and enstatite + albite + H_2O at any specific pressure is displaced to only slightly lower temperatures compared to the corresponding equilibrium for the undersaturated system. For a more typical effect, the reader is referred to dehydration curves

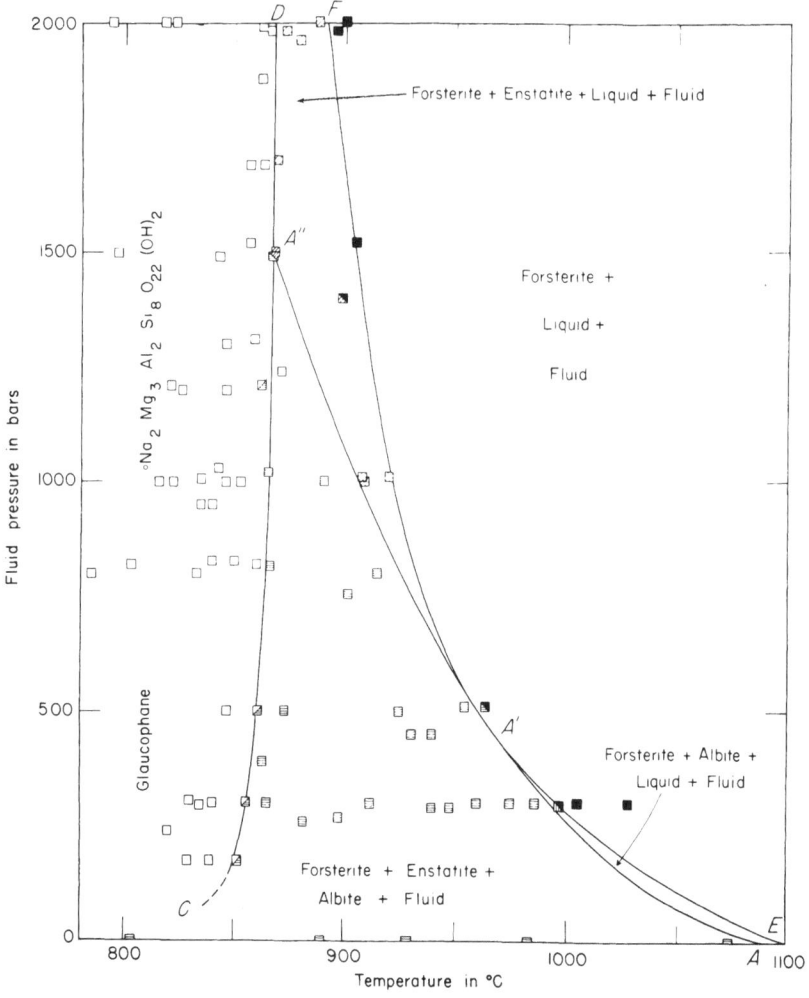

Figure 42

Experimentally determined phase relations for the glaucophane bulk composition (Ernst, 1961). Reactions involving glaucophane I were reversed at 820 and 2000 bars fluid pressure. Divided run symbols indicate persistence of both low and high temperature assemblages of equivalent bulk composition.

published by Evans (1965) for muscovite and for muscovite + quartz.

The low temperature stability limits of glaucophane and glaucophane + quartz were not experimentally located, but Ernst calculated that above about 1000 bars P_{fluid}, reactions such as

glaucophane + H_2O = 2 albite + serpentine,

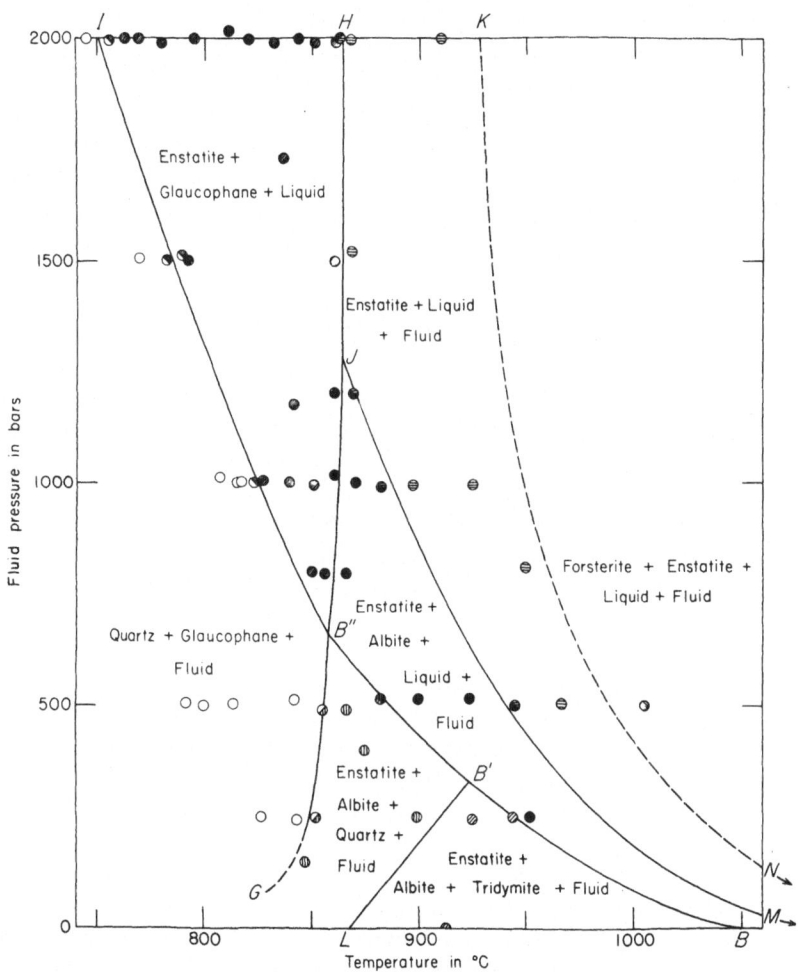

Figure 43

Experimentally determined phase relations for the glaucophane + quartz bulk composition (Ernst, 1961). The reaction involving glaucophane I incongruent melting was reversed at 1500 bars fluid pressure. Divided run symbols indicate persistence of both low and high temperature assemblages of equivalent bulk composition.

glaucophane + 2 quartz = 2 albite + talc,

and 7 glaucophane + 10 quartz = 7 albite + 3 anthophyllite + $4H_2O$,

all involve entropy and volume increases if displaced to the right. Therefore, the glaucophane-bearing assemblage is the low temperature assemblage as well as the experimentally demonstrated high temperature assemblage, being stable at temperatures above those for serpentine, talc, and anthophyllite as shown by Bowen and Tuttle (1949) and Greenwood (1963). Accordingly, associations of albite + serpentine, rarely albite + talc, and albite + anthophyllite should be metastable. In any case, the compatibility of glaucophane + quartz is common and well documented; it needs to be pointed out however, that this natural sodic amphibole contains minor amounts of iron and magnesium, thus only approximates the end-member glaucophane composition.

Magnesioriebeckite, $\circ Na_2Mg_3Fe_2^{+3}Si_8O_{22}(OH)_2$

Stability relations for magnesioriebeckite were determined by Ernst (1960) employing hydrothermal techniques. Because the investigated bulk composition contains iron, relative oxidation states were controlled using suitable oxygen buffers. Amphibole dehydration curves for the bulk composition $Na_2O \cdot 3MgO \cdot 2FeO_x \cdot 8SiO_2$ + excess H_2O are presented in Figs. 44 and 45, conventional P_{fluid}-T diagrams employing the hematite-magnetite and magnetite-wüstite buffers respectively. The value of x ranged from approximately 1.3 to 1.5 depending on physical conditions of the experiments. Oxide mixtures were used in nearly all experiments, but results are entirely compatible with reaction reversals (equilibrium was demonstrated by employing synthetic crystalline charges for at least one isobar for each magnesioriebeckite dehydration curve). The thermal stability limit of magnesioriebeckite approximates that of glaucophane, as shown in the figures. As a matter of fact, substitution of Fe^{+3} for Al apparently elevates the amphibole dehydration temperature more than 60 C° in this case.

At 1000 bars fluid pressure, with oxygen fugacity defined by the hematite-magnetite buffer, magnesioriebeckite melts incongruently at 928 ± 5° C (Fig. 44). The high temperature assemblage of anhydrous solids + melt contains abundant ferrous iron, so reduced f_{O_2} favors this assemblage over the predominantly ferric iron-bearing sodic amphibole; accordingly, with f_{O_2} defined by the magnetite-wüstite buffer, the amphibole melts incongruently at 894 ± 5° C at 1000 bars fluid pressure (Fig. 45). Fig. 46 illustrates the effect of oxygen fugacity on phase equilibria. Subsequent x-ray work (ERNST, 1963a) has indicated that under more reducing conditions the amphibole becomes more ferrous. Thus near the f_{O_2}-T base of

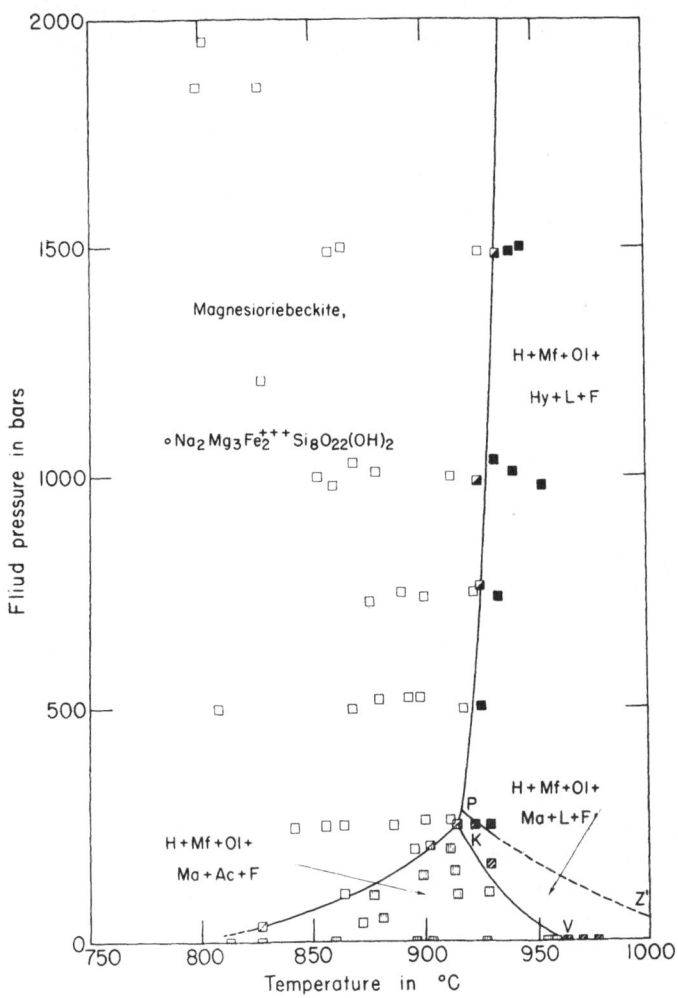

Figure 44

Experimentally determined phase relations for the magnesioriebeckite bulk composition, with oxygen fugacities defined by the hematite-magnetite buffer (Ernst, 1960). Equilibrium for the incongruent melting of magnesioriebeckite was demonstrated at 750 bars fluid pressure. Divided run symbols indicate persistence of low and high temperature assemblages of equivalent bulk composition. Abbreviations are as follows: H = hematite; Mf = magnesioferrite; Mw = magnesiowustite; Ol = olivine; Hy = hypersthene; Ma = $Na_2O \cdot 5(Mg,Fe)O \cdot 12SiO_2$; N = $Na_2O \cdot 2(Mg,Fe)O \cdot 6SiO_2$; Ac = acmite; L = liquid; F = fluid.

the magnetite field, the sodic amphibole represents solid solution between magnesioriebeckite and magnesioarfvedsonite.

Because of the presence of several phases undersaturated with respect to SiO_2 in the high temperature assemblage of equivalent bulk composition, addition of silica would lower the thermal stability range of this

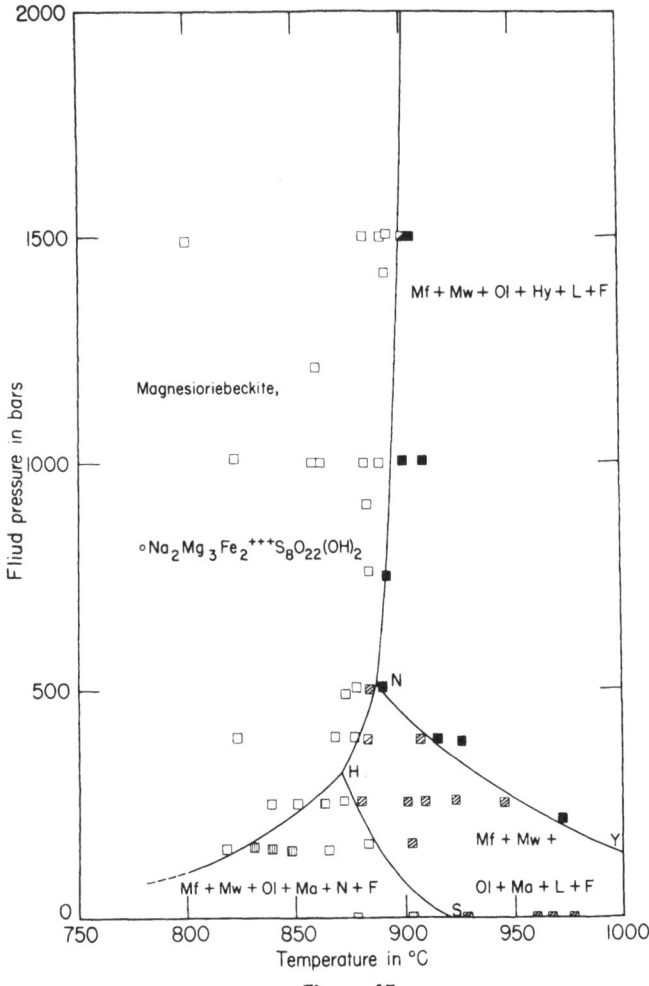

Figure 45

Experimentally determined phase relations for the magnesioriebeckite bulk composition, with oxygen fugacities defined by the magnetite-wüstite buffer (Ernst, 1960). Equilibrium for the incongruent melting of magnesioriebeckite was demonstrated between 750–900 bars fluid pressure. Abbreviations are identical to those of Fig. 44.

amphibole. Neither this, nor the low temperature magnesioriebeckite stability limit (if any exists) was investigated.

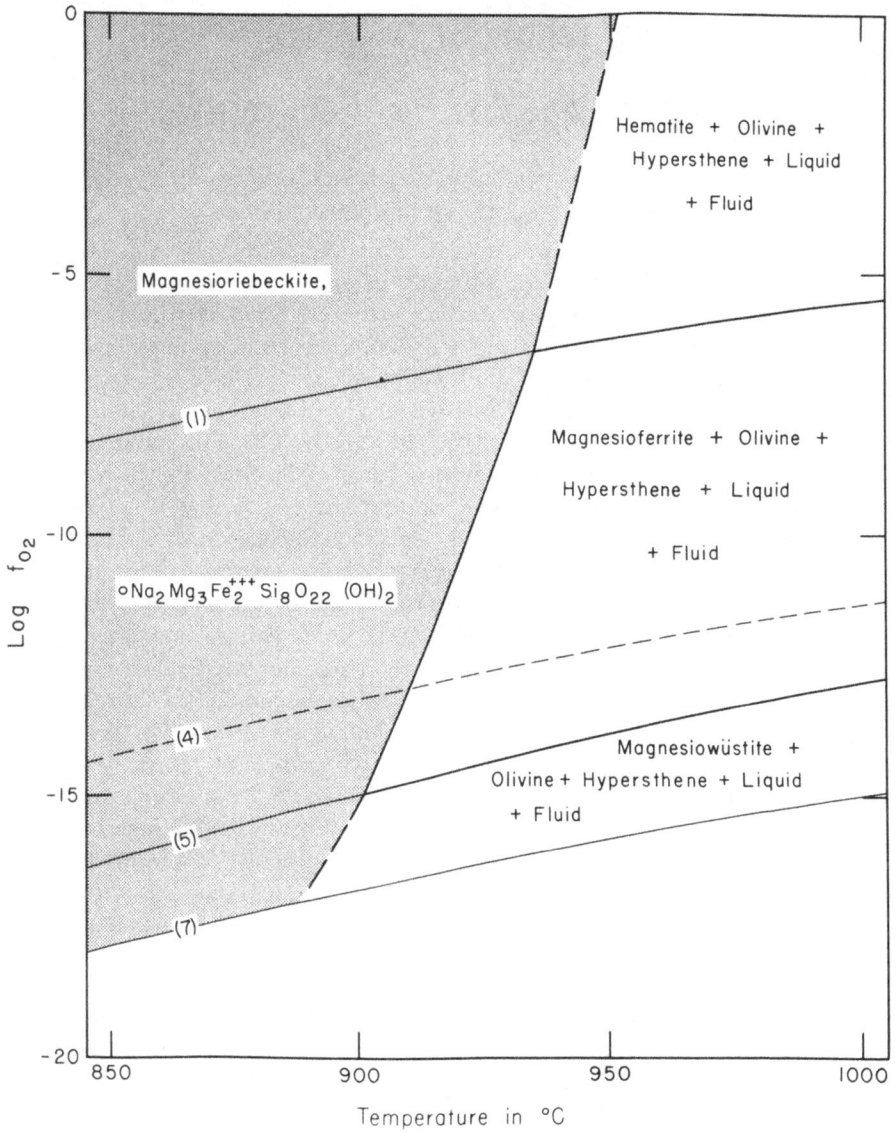

Figure 46

Isobaric log f_{O_2}-T diagram for the magnesioriebeckite bulk composition at 2000 bars fluid pressure (Ernst, 1960). Field boundaries are dashed where inferred. Buffer curves (1), (4), (5) and (7) correspond to those of Fig. 17.

Riebeckite and Riebeckite-Arfvedsonite Solid Solutions,

$$\circ Na_2 Fe_3^{+2} Fe_2^{+3} Si_8 O_{22}(OH)_2 -$$

$$Na_{2.4} Fe_{4.9}^{+2} Fe_{0.7}^{+3} Si_{7.7} Fe_{0.3}^{+3} O_{22}(OH)_2$$

Stable phase relations for the bulk composition $Na_2O \cdot 5FeO_x \cdot SiO_2 +$ H_2O under hydrothermal conditions have been determined by ERNST (1962); x varied from about 1.1 to 1.2, depending on the oxygen fugacity, temperature and fluid pressure. Mixtures of oxides and synthetic crystalline assemblages, employed as starting materials, gave identical results; reaction reversals demonstrated equilibria involving amphibole, even at temperatures as low as 500° C. The f_{O_2} range investigated was bounded by the hematite-magnetite and iron-wüstite or iron-magnetite equilibria. Fine-grained blue riebeckite, $\circ Na_2 Fe_3^{+2} Fe_2^{+3} Si_8 O_{22}(OH)_2$, is stable under relatively oxidizing conditions where f_{O_2} is defined by the hematite-magnetite buffer, but only at low temperatures, i.e., below 496 + 5° C at 1000 bars fluid pressure (Fig. 47). Under these conditions, the charge can be crystallized to essentially 100 percent amphibole. With more reducing conditions, the high temperature stability limit of the amphibole is elevated and the crystals become coarser grained and greenish, reflecting solid solution towards the more ferrous amphibole, arfvedsonite (Fig. 48). Small amounts (approximately 5 percent) quartz coexist stably with the amphibole. This expansion of the amphibole stability field to higher temperatures occurs because the high temperature condensed assemblage, quartz + acmite + magnetite and/or fayalite is more oxidized than riebeckite-arfvedsonite solid solution. Where oxygen fugacity is defined by the wüstite-iron buffer, green amphibole coexisting with about 10–15 percent quartz, has attained the approximate composition $Na_{2.4} Fe_{4.9}^{+2} Fe_{0.7}^{+3} Si_{7.7} Fe_{0.3}^{+3} O_{22}(OH)_2$, as inferred from micrometric analyses. Its maximum thermal stability occurs at 695 ± 5° C at 1000 bars P_{fluid}, and above 1200 bars riebeckite-arfvedsonite melts incongruently as shown in Fig. 49. The dependence of the amphibole thermal stability limit on f_{O_2} is illustrated in Fig. 50.

The variations of composition as indicated by point counts, N_z and unit cell volumes of riebeckite-arfvedsonite solid solutions are presented in Figs. 51, 52 and 53 respectively. With decreasing oxygen fugacity and/or increasing temperature, the silicon proportion of the amphibole, $Si/Na + Fe + Si$, diminishes as the iron becomes more reduced. Apparently the A structural sites which are vacant in riebeckite, become at first partially, then more fully occupied by Na; the rest of the iron and sodium redistribute themselves among the M positions. Thus the composition of the amphibole is a function of f_{O_2} and T; the Na/Fe ratio remains constant, but Si/Na+Fe+Si declines with decreasing oxidation state. The A site is filled completely near the f_{O_2}-T base of the magnetite field. Further reduction in the silicon proportion of the amphibole at even lower oxidation states which is suggested by micrometric analyses only can be accomplished through substitution of minor tetrahedrally coordinated ferric iron for Si; continued reduction of the Fe^{+3}/Fe^{+2} ratio would seem to necessitate more extensive hydration. As shown in Fig. 53, the unit cell volume increases abruptly by more than a percent as

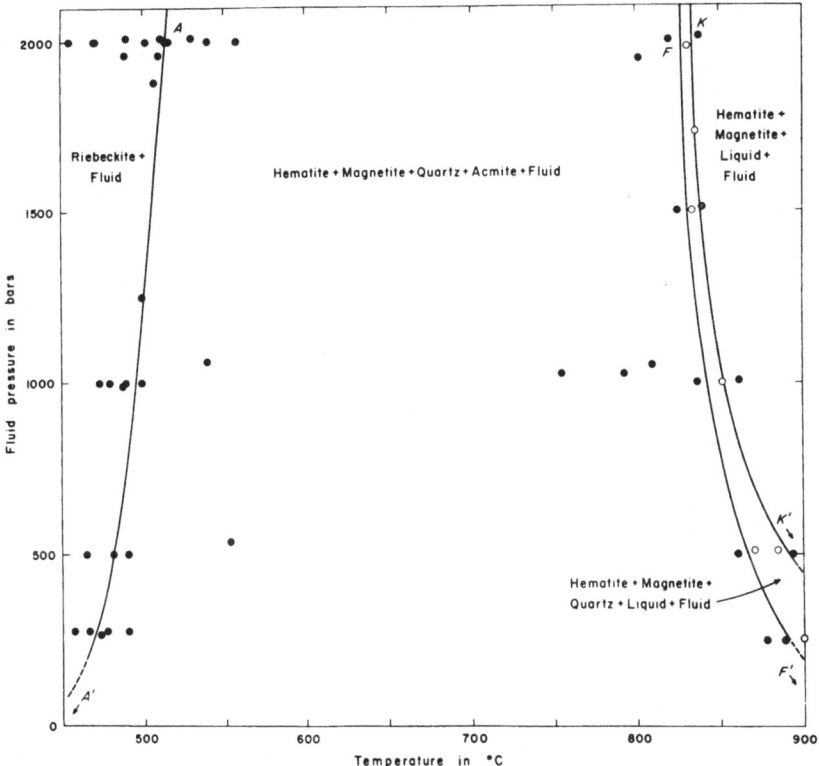

Figure 47

Experimentally determined phase relations for the riebeckite bulk composition, with oxygen fugacities defined by the hematite-magnetite buffer (Ernst, 1962). Divided run symbols indicate persistence of both low and high temperature assemblages of equivalent bulk composition. Reversal of the reaction involving decomposition of riebeckite was demonstrated at 2000 bars fluid pressure.

small amounts (about 0.3 ions per 8 four-fold positions) of ferric iron enter tetrahedral sites. The volume change is accounted for partly by increase in the length of the b axis, but principally by increased separation of parallel double-chains in the direction $a \sin \beta$. As shown by the data

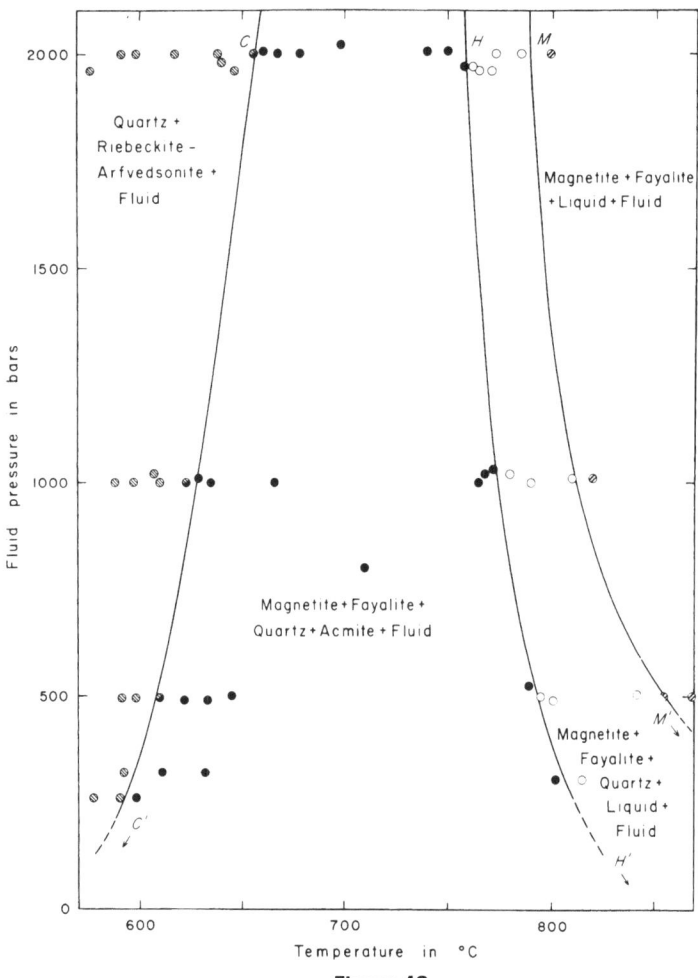

Figure 48

Experimentally determined phase relations for the riebeckite bulk composition, with oxygen fugacities defined by the magnetite + quartz − fayalite buffer (Ernst, 1962). Divided run symbols indicate persistence of both low and high temperature assemblages of equivalent bulk composition. Reversal of the reaction involving decomposition of riebeckite-arfvedsonite was demonstrated at 2000 bars fluid pressure.

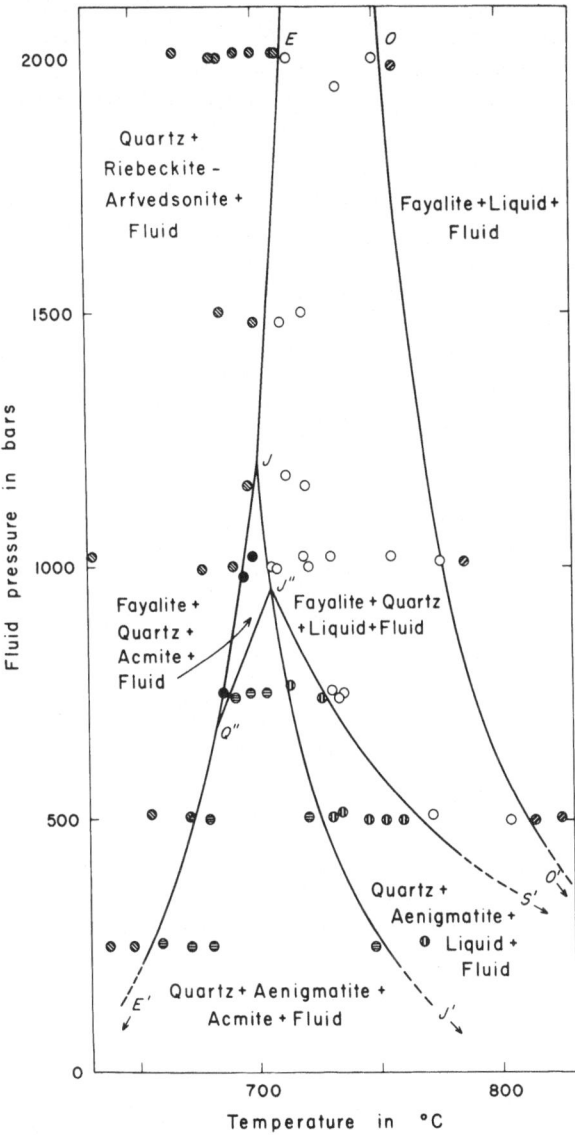

Figure 49

Experimentally determined phase relations for the riebeckite bulk composition, with oxygen fugacities defined by the wüstite-iron buffer (Ernst, 1962). Divided run symbol indicates persistence of both low and high temperature assemblages of equivalent bulk composition. Reversal of the reaction involving incongruent melting of riebeckite-arfvedsonite was demonstrated at 1500 and 2000 bars fluid pressure.

of Table 2, the change from riebeckite to riebeckite-arfvedsonite solid solution involves Δb increment of 0.09 Å (=0.5 percent) and Δa increment of 0.12 Å (=1.2 percent). Evidently accommodation of the large ferric ions in four-fold sites results in considerable inflation of the double-chains parallel to a sin β. Refringence values scarcely are affected by the compositional variation, probably because the unit cell volume increases proportionately as the high atomic number cations are added—thus electron density remains nearly constant.

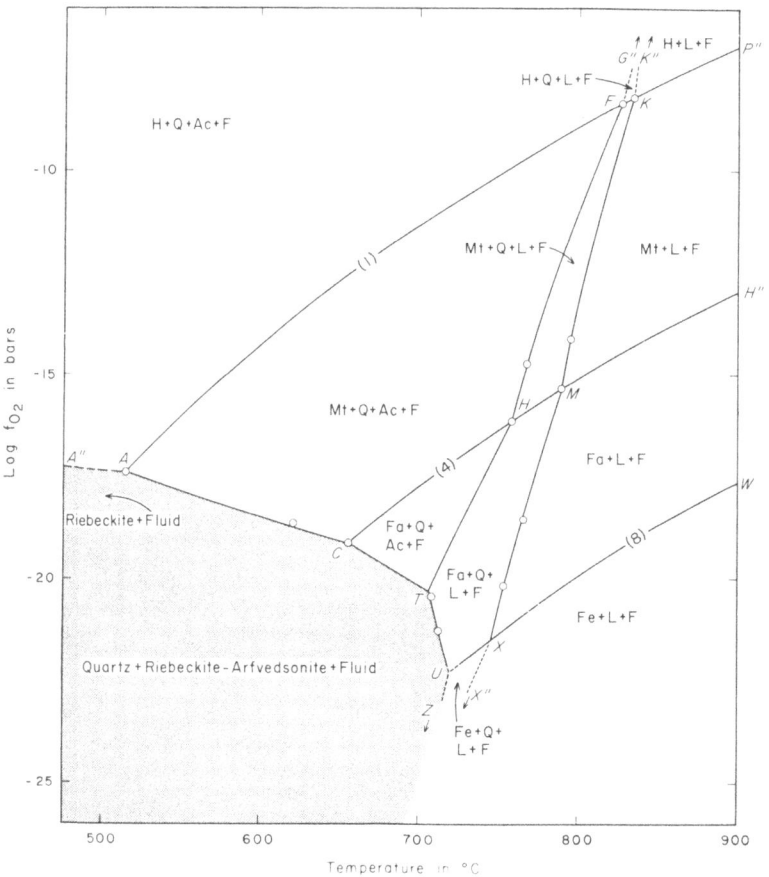

Figure 50

Isobaric log f_{O_2}-T diagram for the riebeckite bulk composition at 2000 bars fluid pressure (Ernst, 1962). Field boundaries are dashed where calculated or in-ferred. Field boundaries (1), (4) and (8) coincide with buffer curves of the same numbers shown in Fig. 17. Abbreviations are: H = hematite; Mt = magnetite; Ac = acmite; Q = quartz; Fa = fayalite; Fe = iron; L = liquid; F = fluid.

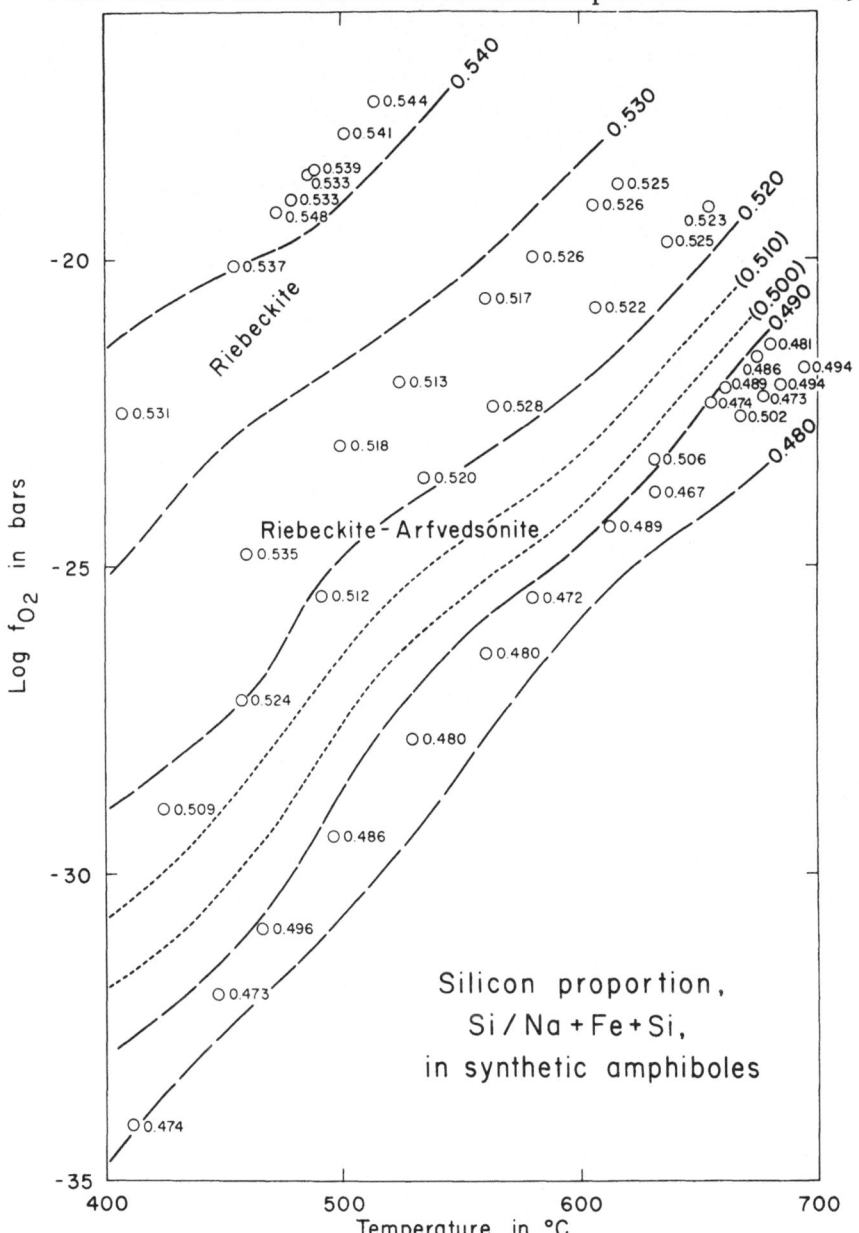

Figure 51

Variation in silicon proportion of synthetic riebeckite-arfvedsonite as a function of f$_{O_2}$ and T (Ernst, 1962). Total pressure variation does not affect the composition of the amphibole.

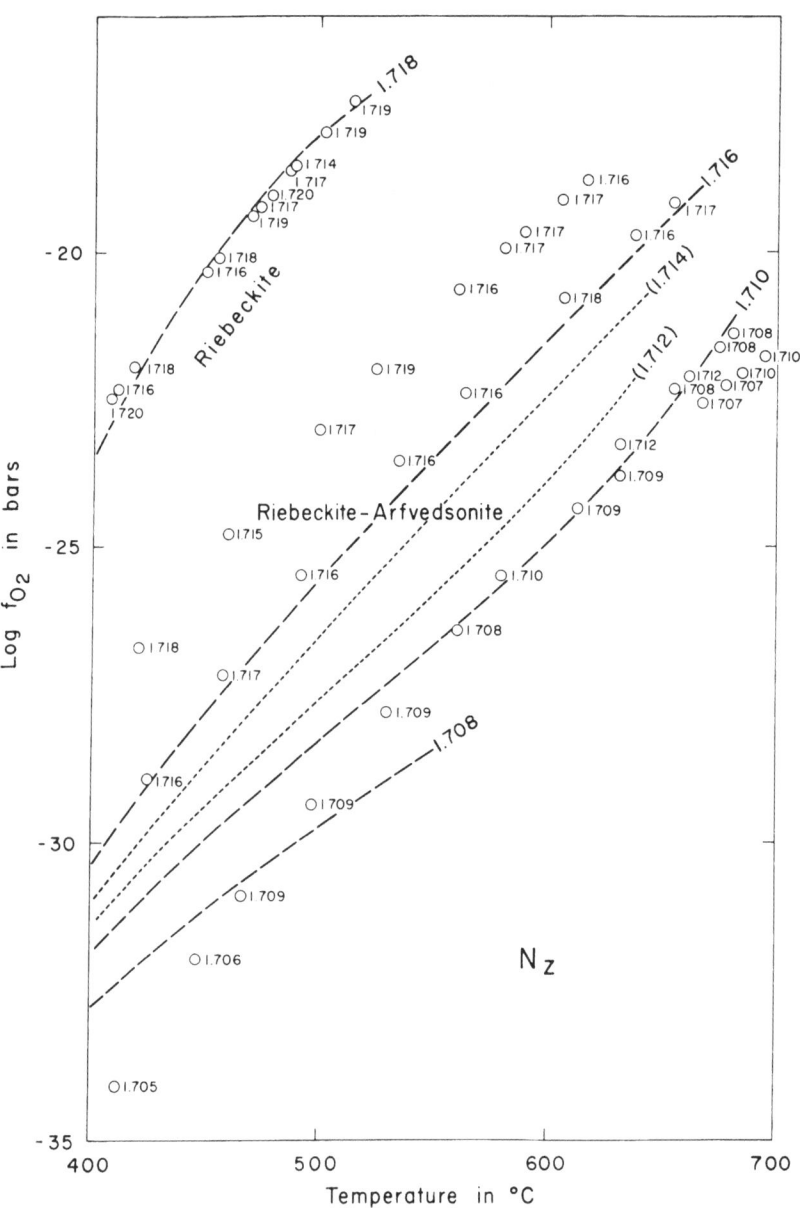

Figure 52

Variation of N_z of synthetic riebeckite-arfvedsonite as a function of log f_{O_2} and T (Ernst, 1962). Refringence decrease is directly related to diminution of the silicon proportion in the amphibole, as may be seen by comparison of this figure with Fig. 51.

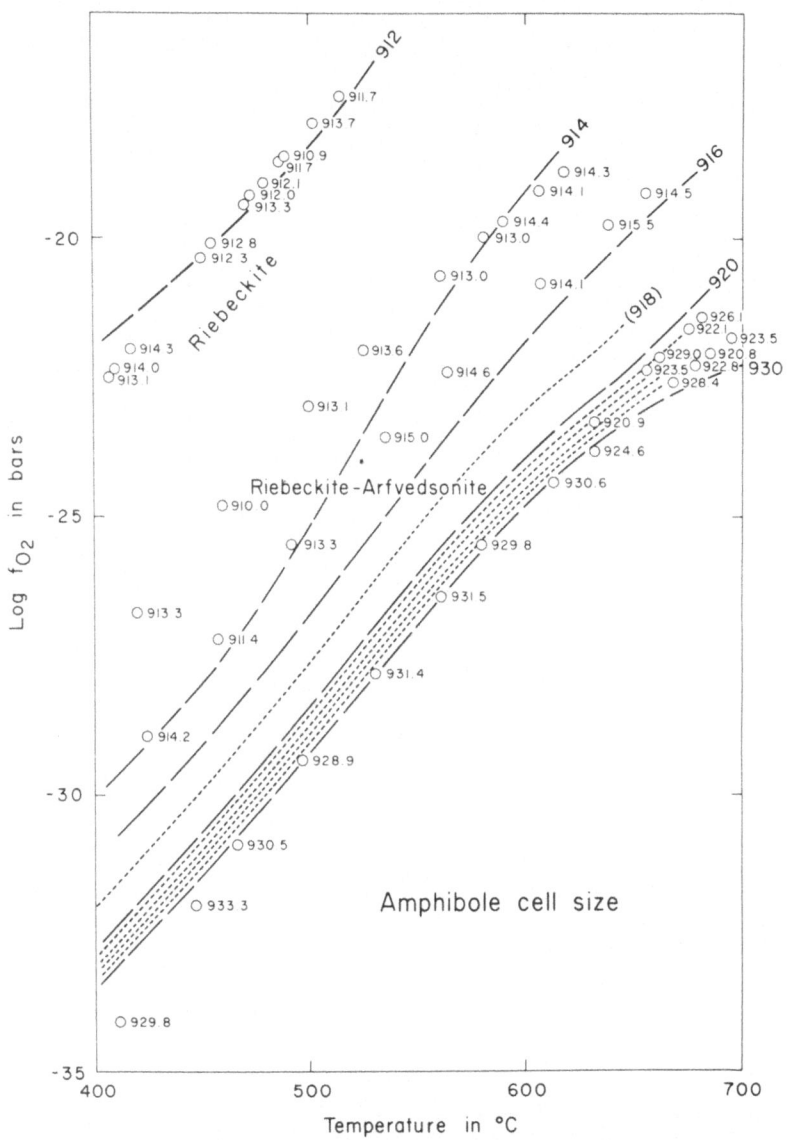

Figure 53

Variation in unit cell volume of synthetic riebeckite-arfvedsonite as a function of log f_{O_2} and T (Ernst, 1962). Cell volume increase is directly related to the decrease in silicon proportion in the amphibole, as may be seen by comparison of this figure with Fig. 51. Note that the unit cell volume increases precipitously at low oxygen fugacities where small amounts of Fe^{+3} are inferred to replace Si in tetrahedral coordination.

As clearly indicated in Fig. 54, at appropriate oxidation states the high temperature stability limit of riebeckite is more than 400 C° lower than that of magnesioriebeckite.

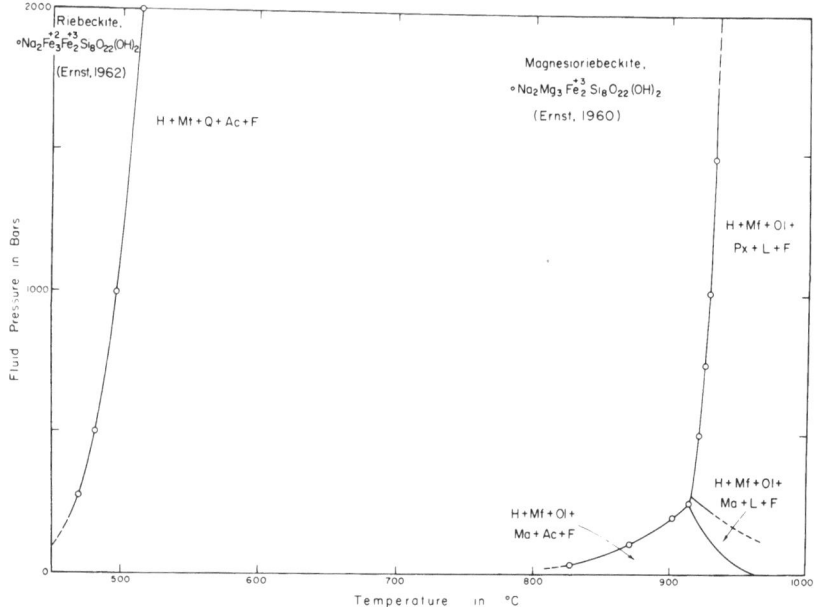

Figure 54

Comparison of thermal stability limits of magnesioriebeckite (Ernst, 1960) and riebeckite (Ernst, 1962). In both cases, f_{O_2} is defined by the hematite-magnetite buffer. Abbreviations are the same as those of Figs. 44 and 50.

Ferroglaucophane, $\circ Na_2Fe_3^{+2}Al_2Si_8O_{22}(OH)_2$

The stable synthesis of ferroglaucophane has not been accomplished. At low pressure the (large volume) equilibrium assemblage appears to consist of albite + fayalite + magnetite + fluid, or albite + magnetite + quartz, depending on the oxygen fugacity. At pressures in excess of about 25 kb, sodic clinopyroxene + garnet + quartz + fluid (small volume) seem to crystallize from the ferroglaucophane bulk composition. Ferroglaucophane might have a stability field at temperatures and pressures intermediate between these two assemblages, but such relations are speculative at present.

Polymorphism in Sodic Amphiboles

The unit cell volume of synthetic glaucophane produced at less than 2000 bars is two percent greater than those of natural analogues close to the end-member composition, as previously mentioned. This discrepancy encouraged further high pressure experimentation (ERNST, 1963a). Glaucophane synthesized at elevated pressures exhibits lattice parameters virtually indistinguishable from those of natural samples. Phase relations were determined subsequently for the bulk compositions glaucophane, magnesioriebeckite, riebeckite, $Gl_{75}Rieb_{25}$, $Gl_{50}Rieb_{50}$ (crossite) and

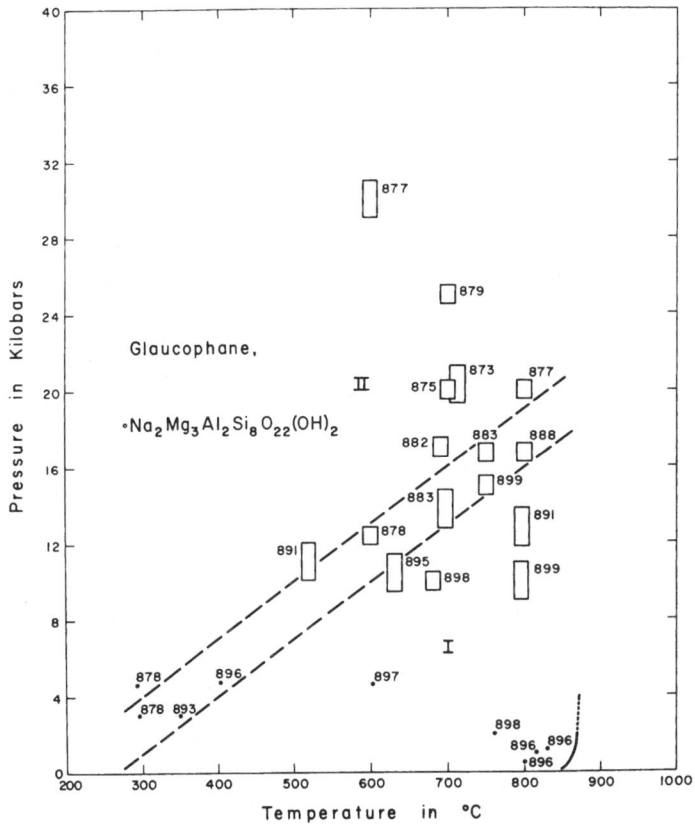

Figure 55

Experimentally determined variation of synthetic glaucophane unit cell volume with temperature and total pressure (Ernst, 1963a). Size of run symbol indicates P-T uncertainty; values for unit cell volumes are given in cubic angstroms. Reversal of reaction has been demonstrated both at about 300–350° C and about 700–800° C.

$Gl_{25}Rieb_{75}$. Hydrothermal equipment was employed at pressures less than 5000 bars, piston-cylinder (BOYD and ENGLAND, 1960) and piston-anvil devices (GRIGGS and KENNEDY, 1956) at pressures in excess of 10 kb; where necessary, oxygen fugacities were controlled by the hematite-magnetite buffer.

The transition between the large volume polymorph of $^\circ Na_2Mg_3Al_2$-$Si_8O_{22}(OH)_2$, designated glaucophane I, and the small volume polymorph, glaucophane II, appears to be gradational and reversible; intermediate cell volumes are restricted to a zone about three kb wide, the

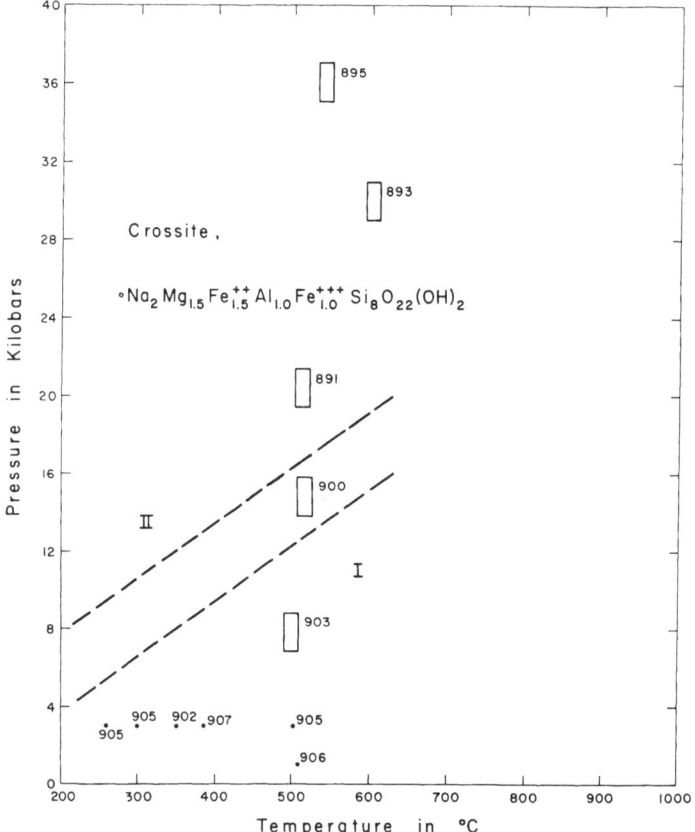

Figure 56

Experimentally determined variation of synthetic crossite unit cell volume with temperature and total pressure (Ernst, 1963a). Oxygen fugacities are defined by the hematite-magnetite buffer. Size of run symbol indicates P-T uncertainty; values for unit cell volumes are given in cubic angstroms. Reversal of reaction has been demonstrated at about 500–600° C.

midpoints of which are defined by the equation T = 220 + 33P, where T = °C and P = kb (see Fig. 55). The transition apparently is second order. Volume increment is a function mainly of increase in the b axis repeat, as discussed in Chapter II. Riebeckite and magnesioriebeckite do not exhibit this variation in cell volume, but crossite shows a one percent decrease in volume at elevated pressure as shown in Fig. 56. The magnitude of ΔV apparently is proportional to the octahedrally coordinated aluminum in the amphibole. Reasoning by analogy with the structural determination of magnesioriebeckite by WHITTAKER (1949), which demonstrated Fe^{+3} (and Al) concentrated in M_2, it was inferred (ERNST, 1963a; COLVILLE et al., 1966) that aluminum was more ordered in the M_2 site in glaucophane II and more randomly distributed among M_1, M_2 and M_3 sites in glaucophane I. The recent three-dimensional structural refinement of natural glaucophane II by PAPIKE and CLARK (in press) substantially demonstrates such ordering in M_2 for a well-crystallized low temperature sodic amphibole.

Richterite, $Na_2CaMg_5Si_8O_{22}(OH)_2$, and Eckermannite, $Na_3Mg_4AlSi_8O_{22}(OH)_2$

These two amphiboles have been synthesized hydrothermally by PHILLIPS and ROWBOTHAM (1967). In neither study was equilibrium demonstrated. Richterite crystallized from the oxide mixture $Na_2O \cdot CaO \cdot 5MgO \cdot 8SiO_2$ + excess H_2O between 750–1000° C at 1000 bars fluid pressure. The thermal stability limit apparently was not exceeded, thus lending support to the suggestion put forth in Chapter III that richterites may represent aluminum-poor high temperature hypersolvus Ca-Na amphiboles.

Synthetic eckermannite was crystallized from the oxide mixture of bulk composition $1.5Na_2O \cdot 4MgO \cdot 0.5Al_2O_3 \cdot 8SiO_2$ + excess H_2O, but was always accompanied by a talc-like phase interpreted as metastable by PHILLIPS and ROWBOTHAM. Synthetic eckermannite also seems to be quite refractory, having been produced in the 800–1000° C range at 1000 bars fluid pressure.

"Magnesiorichterite," $Na_2Mg_6Si_8O_{22}(OH)_2$, and "Fluormagnesiorichterite," $Na_2Mg_6Si_8O_{22}F_2$

CHRISTOPHE-MICHEL-LÉVY (1957) synthesized an amphibole from an initial mixture consisting of $CaCO_3$, $MgCO_3$, $NaCO_3$, SiO_2 and H_2O at about 450° C; the value of P_{fluid} was not stated, but judging from the other experiments reported was about 300–400 bars. This author indicated that the composition of the amphibole obtained was $Na_{1.5}Ca_{0.7}Mg_{5.4}Si_8O_{20}$-$(OH)_{3.8}$. The synthetic double-chain silicate is thus compositionally intermediate between richterite and "magnesiorichterite." Provided the water analysis is correct, protons are not confined to the O_3 structural position in this amphibole; perhaps some H^+ ions are related to other non-bridging anion sites. Only minor amounts of the excess could be accommodated as H_3O^+ in the A position, which is partially filled by sodium. Reaction reversal was not demonstrated.

SIEFERT and SCHREYER (personal communication) produced "magnesiorichterite" from glass $\pm$ oxide mixtures of the requisite bulk composition, $Na_2O \cdot 6MgO \cdot 8SiO_2$ + excess H_2O. This synthetic amphibole melts incongruently at $965 \pm 20°$ C at 1000 bars fluid pressure to forsterite, an osumilite-type phase (Na_2O-$2MgO$-$6SiO_2$) and liquid. It is also noteworthy that in addition to "magnesiorichterite", other apparently more hydrous amphiboles within the system Na_2O-MgO-SiO_2-H_2O have been synthesized (GIER et al., 1964; SEIFERT and SCHREYER, personal communication). The fluorine analogue of this species has been prepared by GIBBS et al. (1962), both from melts and by recrystallization of oxide mixtures.

Natural Occurrences of the Sodic Amphiboles

The Al-rich sodic amphiboles are practically confined to parageneses appropriate to blueschist (and transitional blueschist-greenschist) metamorphic facies. In contrast, the iron-rich sodic amphiboles occur in alkalic igneous rocks and gneisses, as well as in lower grade metamorphic rocks, and even as authigenic minerals. The greater variety of geologic environments suitable for the iron-rich sodic amphiboles compared to aluminous types results not from more extensive thermal stability ranges (as shown by experimental investigations, e.g., compare Figs. 42, 44 and 49), but rather from the fact that appropriate chemical conditions are required. Very few rock bulk compositions are high in alkalis, alumina and magnesia, and simultaneously low in lime and iron, thus accounting for the absence of glaucophane I in nature; occurrences of peralkalinity and iron enrichment however are not as rare.

Although glaucophane should be stable over a very wide range of physical conditions in Na_2O+MgO-rich, CaO-poor compositional environments, analyses of glaucophane schists do not indicate a systematic difference in these or any other oxides compared to greenschist and epidote-amphibolite analyses (ERNST, 1963b). It should be noted that most glaucophanes are compositionally intermediate between Al- and iron-rich end-members, hence properly should be termed crossites. Inasmuch as blueschists do not have peculiar rock bulk compositions compared to the other more "normal" metamorphic facies types, the pressure-temperature conditions during recrystallization must have been responsible for the phase assemblage. Therefore such glaucophane schists belong to a separate metamorphic facies, employing the concept as defined by ESKOLA (1929).

Many of the minerals of glaucophane schists also appear commonly in greenschists and in albite-epidote amphibolites, but jadeitic pyroxene (DE ROEVER, 1955; BLOXAM, 1956, 1959, 1960; SEKI, 1960; McKEE, 1962a), metamorphic aragonite (COLEMAN and LEE, 1962; McKEE, 1962b; ERNST, 1965; GHENT, 1965), lawsonite (DE ROEVER, 1947; SEKI, 1958; McKEE, 1962a) and glaucophane II (ERNST, 1963a) are virtually confined to blueschist facies parageneses. Jadeite (ROBERTSON et al., 1957; BIRCH and LeCOMTE, 1960) and aragonite (CLARK, 1957) are definitely high pressure minerals, and in the presence of excess quartz, so is lawsonite (CRAWFORD and FYFE, 1965). Glaucophane I has never been reported, presumably because the metamorphic geothermal gradient lies well within the stability field for polymorph II. We can conclude that physical conditions involved relatively high pressures and low temperatures; where active, metasomatism merely would have changed the proportions of the individual phases in the final equilibrium assemblage.

Typical mineral parageneses from the California Coast Ranges and the outer metamorphic belt of Japan (ERNST and SEKI, 1967) are presented in Figs. 57, 58 and 59. The *in situ* California mafic blueschists are typified by the association of glaucophane-crossite + lawsonite ± aragonite; jadeitic pyroxene + quartz + lawsonite ± aragonite occur in the more intensely recrystallized metagraywackes. On the other hand, Shikoku mafic schists contain coexisting albite + actinolite + crossite + epidote + calcite; metaclastic rocks carry albite + calcite + quartz ± epidote. Inferred physical conditions based on experimental phase equilibrium and oxygen isotopic data are: California, 200–300° C, 6.5–8 kb; Japan, 250–400° C, 5–7 kb.

Authigenic riebeckite-magnesioriebeckite has been described from the Green River Formation of Utah by MILTON and EUGSTER (1959). This mineral occurrence undoubtedly reflects the former presence of a sodium-rich pore fluid, as testified also by the abundance of other Na-silicates. More widespread examples of the riebeckite-magnesioriebeckite series are to be found in the meta-ironstones of South Africa, Western Australia and Bolivia (PEACOCK, 1928; HALL, 1930; MILES, 1942; AHLFELD, 1943; DE VILLIERS, 1949; WYMOND and WILSON, 1951). As also concluded from study of the authigenic Na-amphibole, most authors agree that certain layers of the original iron formations (including interstratal solutions) were peralkaline, hence compositions were appropriate for crystallization of members of the riebeckite-magnesioriebeckite series. Fibrous varieties, known collectively as crocidolite, are of commercial value as asbestos. Relatively high oxidation states are indicated by the coexistence of crocidolite with hematite; judging from the feeble nature of the recrystallization, temperatures must have been low.

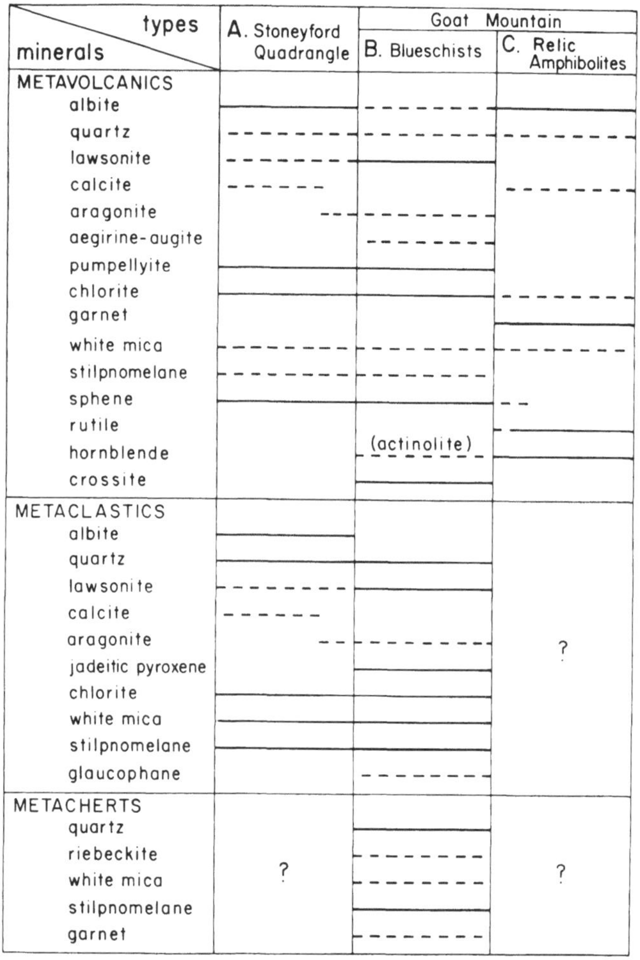

types / minerals	A. Stoneyford Quadrangle	Goat Mountain B. Blueschists	C. Relic Amphibolites
METAVOLCANICS			
albite			
quartz			
lawsonite			
calcite			
aragonite			
aegirine-augite			
pumpellyite			
chlorite			
garnet			
white mica			
stilpnomelane			
sphene			
rutile			
hornblende		(actinolite)	
crossite			
METACLASTICS			
albite			
quartz			
lawsonite			
calcite			
aragonite			?
jadeitic pyroxene			
chlorite			
white mica			
stilpnomelane			
glaucophane			
METACHERTS			
quartz			
riebeckite	?		?
white mica			
stilpnomelane			
garnet			

Figure 57

Mineral parageneses in progressive metamorphism of the Franciscan terrane, northern California (Coleman and Lee, 1963; Ghent, 1965; Ernst and Seki, 1967). Assemblage C is partly converted to assemblage B.

Low grade riebeckite-aegirine greenschists have been described by WHITE (1962) from New Zealand, and higher grade riebeckite gneisses from Spain by FLOOR (1966). In these cases too, the rocks have $(Na_2O + K_2O)/Al_2O_3$ ratios greater than unity, and are relatively oxidized, in harmony with the experimental data.

Riebeckite of essentially the end-member composition was reported from Quincy, Massachusetts, as a primary pegmatite phase associated

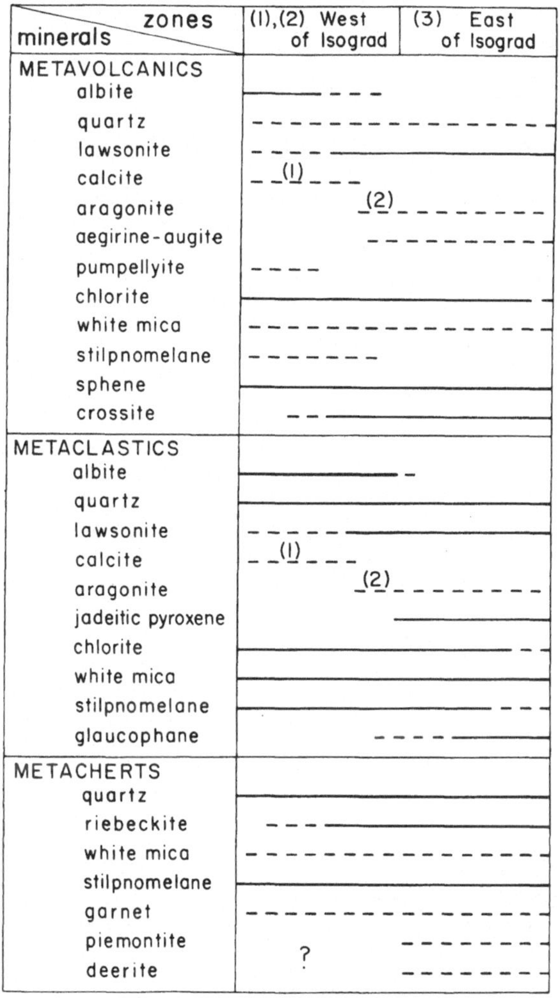

minerals / zones	(1),(2) West of Isograd	(3) East of Isograd
METAVOLCANICS		
albite		
quartz		
lawsonite		
calcite	(1)	
aragonite	(2)	
aegirine-augite		
pumpellyite		
chlorite		
white mica		
stilpnomelane		
sphene		
crossite		
METACLASTICS		
albite		
quartz		
lawsonite		
calcite	(1)	
aragonite	(2)	
jadeitic pyroxene		
chlorite		
white mica		
stilpnomelane		
glaucophane		
METACHERTS		
quartz		
riebeckite		
white mica		
stilpnomelane		
garnet		
piemontite	?	
deerite		

Figure 58

Mineral parageneses in progressive metamorphism of the Franciscan terrane, central California (McKee, 1962a, b; Ernst, 1965; Ernst and Seki, 1967).

Amphiboles

with quartz + alkali feldspars + aegirine + hematite and possible magnetite (PALACHE and WARREN, 1911). Assuming fluid pressure was equal to lithostatic pressure, the temperature of crystallization must have approximated 500 ± 50° C, depending on the depth of emplacement. Other sodic amphiboles in granites and syenites from, *e.g.*, Vermont, Nigeria, Greenland and Sakhalin (CHAPMAN and WILLIAMS, 1935; MACKAY *et al.*, 1949; YAGI, 1953; WOODARD, 1957; JACOBSON *et al.*, 1958; UPTON,

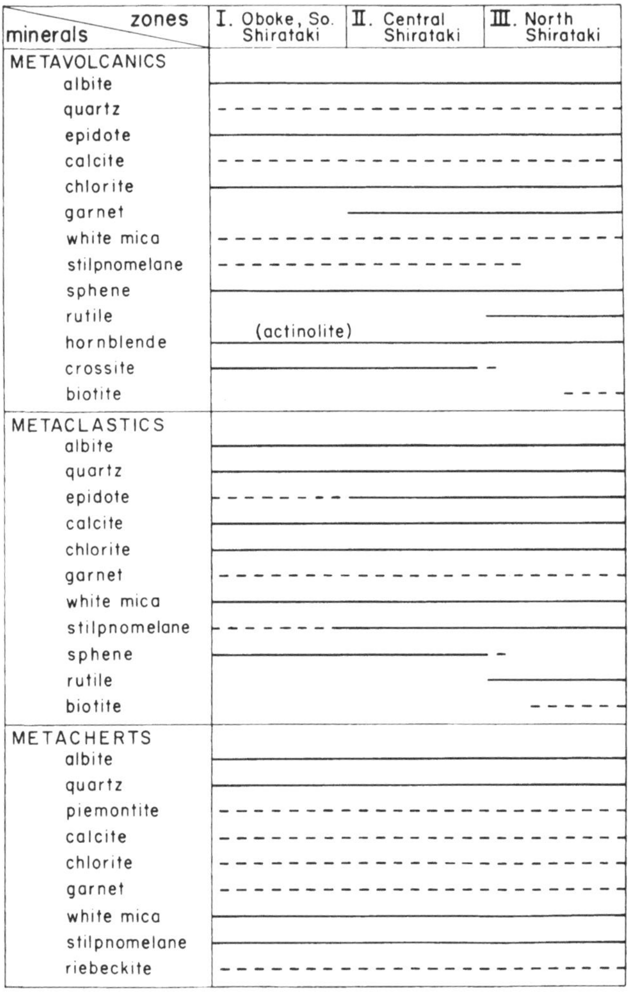

minerals \ zones	I. Oboke, So. Shirataki	II. Central Shirataki	III. North Shirataki
METAVOLCANICS			
albite			
quartz			
epidote			
calcite			
chlorite			
garnet			
white mica			
stilpnomelane			
sphene			
rutile			
hornblende	(actinolite)		
crossite			
biotite			
METACLASTICS			
albite			
quartz			
epidote			
calcite			
chlorite			
garnet			
white mica			
stilpnomelane			
sphene			
rutile			
biotite			
METACHERTS			
albite			
quartz			
piemontite			
calcite			
chlorite			
garnet			
white mica			
stilpnomelane			
riebeckite			

Figure 59

Mineral parageneses in progressive metamorphism of the Sanbagawa terrane, central Shikoku (Hide, 1961; Banno, 1964; Ernst and Seki, 1967).

1960; BORLEY, 1963) are either arfvedsonite, fluorarfvedsonite or fluor-riebeckite. The replacement of hydroxyl by fluorine greatly elevates the high temperature stability limit of amphibole, thus accounting for the stabilization of fluorriebeckite at magmatic temperatures. As shown by the experimental study (see Fig. 50), at relatively low oxidation states riebeckite-arfvedsonite solid solution also is stable at magmatic temperatures, even in the absence of fluorine, where it may be associated with acmite, aenigmatite and/or fayalite; all these phases are typical of arfvedsonite-bearing plutons, and indicate close correlation between ex-perimentation and natural occurrence.

Provided we were justified in ignoring the lack of Ti in synthetic aenigmatite, Fig. 49 might be employed to distinguish higher tempera-ture, lower pressure sodic amphibole + aenigmatite-bearing alkalic plutons ($\approx$ hypersolvus granites of TUTTLE and BOWEN, 1958) from lower temperature, higher pressure sodic amphibole + fayalite-bearing ($\approx$ subsolvus) granites. Unfortunately, because natural aenigmatites are enriched in titanium, this reaction, aenigmatite $\rightarrow$ fayalite + acmite, must be applied with considerable caution.

Richterites have been described from alkaline igneous rocks and peg-matites where they typically are found in and adjacent to calc-silicate bodies (LARSEN, 1942; SUNDIUS, 1945); the mineral assemblages are quite diverse. Recently OLSEN (1967) has reported the presence of primary richterite in graphite nodules of an iron meteorite, where it is associated with forsterite, albite and roedderite, an $Na_2O \cdot 5MgO \cdot 12SiO_2$-type min-eral with osumilite structure. Thus the relatively rare soda tremolites evidently are restricted to high temperature environments and to rock bulk compositions rich in both soda and lime relative to alumina. The common association of actinolite with glaucophane or crossite in low grade schists precludes the occurrence of aluminous richterites in such terranes.

Chapter VIII.

CONCLUDING STATEMENT

Comparison of Experimental P-T Curves

The influence of various cationic substitutions, in other words site occupancy, on the P-T ranges of amphiboles may be evaluated from a comparison of the experimentally determined phase relations for pure end-members as shown in Table 3. For instance, tremolite has a thermal stability limit approximately 85–95 C° higher than that of anthophyllite; Ca occupancy of M_4 obviously increases the amphibole high temperature stability limit over the case where Mg occupies this site (however, it must be remembered that the structures of anthophyllite and tremolite differ to some

TABLE 3

INFERRED CATION SITE OCCUPANCY AND THERMAL RANGE
OF SYNTHETIC HYDROXYL-AMPHIBOLE END-MEMBERS

Species	Structural Site Occupancy				High Temperature Stability Limit, °C		
	A	M_4	M_2	$M_1 + M_3$	500 bars	1000 bars	2000 bars
(1) Anthophyllite	°	Mg	Mg	Mg	—	745	765
(2) Pargasite	Na	Ca	Al+Mg	Mg+(Al)	955	1045	—
(3) Ferropargasite	Na	Ca	Al+Fe^{+2}	Fe^{+2}(+Al)	682	800	850
(4) Tremolite	°	Ca	Mg	Mg	800	830	870
(5) Ferrotremolite	°	Ca	Fe^{+2}	Fe^{+2}	437	465	506
(6) Magnesioriebeckite	°	Na	Fe^{+3}(+Mg)	Mg(+Fe^{+3})	921	928	935
(7) Riebeckite	°	Na	Fe^{+3}(+Fe^{+2})	Fe^{+2}(+Fe^{+3})	481	496	515
(8) Glaucophane I	°	Na	Al(+Mg)	Mg(+Al)	859	864	868

(1) GREENWOOD, 1963.
(2) BOYD, 1959. Al + Mg inferred to be disordered.
(3) GILBERT, 1966; ferropargasite phase equilibria, fo_2 defined by the wüstite-iron buffer. Al + Mg inferred to be disordered.
(4) BOYD, 1959.
(5) ERNST, 1966; ferrotremolite phase equilibria, fo_2 defined by the magnetite-iron buffer.
(6) ERNST, 1960; magnesioriebeckite phase equilibria, fo_2 defined by the hematite-magnetite buffer. Fe^{+3} + Mg inferred to be disordered.
(7) ERNST, 1962; riebeckite phase equilibria, fo_2 defined by the hematite-magnetite buffer. Fe^{+3} + Fe^{+2} inferred to be partially disordered.
(8) ERNST, 1961. Al + Mg inferred to be disordered.

extent). Magnesioriebeckite is stable to temperatures approximately 60–70 C° in excess of the high temperature stability limit of glaucophane I; hence Fe^{+3} occupancy of M_2 increases the amphibole thermal range compared to Al in this position (because of disorder, both amphiboles also contain some Mg in this site). Comparison of the P-T curves for the decomposition of tremolite and of glaucophane reveals that the coupled replacement of Ca by Na in M_4, plus the partial replacement of Mg by Al in M_2 influences the amphibole high temperature stability limit to only a minor extent. From the above discussion we may conclude only tentatively that (1) amphiboles containing Ca or Na in M_4 are somewhat more refractory than those with Mg in this position and (2), amphiboles which carry ferric iron in M_2 are stable to slightly higher temperatures than those containing aluminum and/or magnesium.

From an examination of phase equilibrium data for tremolite, pargasite and magnesioriebeckite and their Fe^{+2}-bearing analogues, it is apparent that the replacement of magnesium by ferrous iron in $M_1 + M_3$ (and M_2?) sites results in a decline of the amphibole thermal stability limit of about 260–430 C°, even under optimum f_{O_2} conditions for stabilization of the ferrous amphibole.

Finally, comparison of pargasite and ferropargasite with tremolite, ferrotremolite and glaucophane stability relations suggests that the substitution of Na for o in the A structural site plus the presence of tetrahedrally coordinated Al replacing Si greatly increases the high temperature stability limit of the amphibole. From the limited amount of evidence currently available, we cannot determine whether this effect is due primarily to occupancy of the A position, to the presence of tetrahedrally coordinated aluminum, or both. In this connection it seems appropriate to mention that arfvedsonite has an increased thermal stability range compared to riebeckite; arfvedsonite, like pargasite and ferropargasite, contains Na in the A site, as well as small amounts of a 4-fold coordinated trivalent ion, Fe^{+3}.

Final Remarks

Our current understanding of amphibole parageneses is exceedingly limited, principally because of the inherent complexities of the problems. The great chemical variations among natural amphiboles increase the difficulties of systematic field and laboratory study. Typically, low reaction rates have impeded experimental amphibole equilibrium studies. Nevertheless, several points now seem clear.

The rarity of ferrous iron-rich natural amphiboles stems from three causes. (1) Fe^{+2} amphiboles have lower thermal stability limits than Mg analogues, even under the most favorable redox conditions. (2) Few rocks which recrystallized at low temperatures also recrystallized at relatively low oxygen fugacities where ferrous amphiboles are stable. (3) Equilibrium element partitioning between a coexisting amphibole and more refractory silicates generally enriches the amphibole in Mg relative to Fe^{+2}.

Six-fold coordinated Fe^{+3} and Al analogues have approximately similar stability ranges, but the ferric iron-bearing amphiboles appear to be more refractory.

The lesser abundances of anthophyllites, cummingtonites and gedrites in nature compared to calcic and sodic amphibole groups stems from two causes. (1) These minerals have rather narrow P-T stability fields, being bounded both at higher temperatures and lower temperatures by other stable assemblages of equivalent bulk compositions. (2) Iron- and magnesium-rich rocks deficient in Ca, Al and alkalis are uncommon.

Sodic amphiboles are less abundant than calcic amphiboles because either peralkaline chemical conditions or relatively high pressures and low temperatures are required for their formation.

Calcic amphiboles are the most common double-chain silicates owing to their broad P-T stability ranges as well as due to the ubiquity of Ca+Fe+Mg+Al-rich rocks.

From experimental studies, crystal structure refinements, Mössbauer and infra-red spectral investigations, increased ordering at low temperatures among amphibole $M_1 + M_3$, M_2 and M_4 sites is indicated. Hence amphibole exchange reaction studies should include consideration of the identification of the participating sites.

Because of the complex and subtle relationships between bulk compositions of the host rocks and amphibole compositions, very little can be said regarding amphibole parageneses in igneous and metamorphic rocks, in spite of painstaking studies by numerous investigators. Hornblendes tend to be pargasitic in mafic and ultramafic intrusives, hastingsitic or arfvedsonitic in felsic and alkalic plutons, thus reflecting both the Fe^{+2}/Mg, Fe^{+3}/Al and Ca/Na ratios of the host and the disparate oxidation states of such bodies. Aluminum contents of low grade metamorphic calcic amphiboles are low, but are high in sodic amphiboles; in the higher grades of regional metamorphism, aluminous Na-bearing Ti-rich calcic amphiboles predominate. High pressures favor both sodic and calcic amphiboles over the Fe-Mg group, and promote a high content of 6-fold coordinated aluminum in the resultant amphibole. More specific statements must await further experimental, analytical and field studies.

References

AHLFELD, F., 1943, Los yacimientos de crocidolita en las yungas de Cochabamba: *Notas Museo La Plata*, **8**, Geol. No. 27, 355–371.

AHRENS, L. H., 1952, The use of ionization potentials. Part I. Ionic radii of the elements: *Geochim. et Cosmochim. Acta*, **2**, 155–169.

AKELLA, J., and WINKLER, H. G. F., 1966, Orthorhombic amphibole in some metamorphic reactions: *Contr. to Min. Pet., Beit. Min. u. Pet.*, **12**, 1–12.

BANCROFT, G. M., BURNS, R. G., and MADDOCK, A. G., 1967, Determination of cation distribution in the cummingtonite-grunerite series: *Am. Mineralogist*, **52**, 1009–1026.

BANNO, S., 1964, Petrologic studies on Sanbagawa crystalline schists in the Bessi-Ino district, central Sikoku, Japan: *Jour. Fac. Sci., Univ. Tokyo*, Sec. II, **15**, 203–319.

BARNES, H. L., and ERNST, W. G., 1963, Ideality and ionization in hydrothermal fluids: the system MgO-H_2O-$NaOH$: *Am. Jour. Sci.*, **261**, 129–150.

BARNES, V. E., 1930, Changes in hornblende at about 800° C: *Am. Mineralogist*, **15**, 393–417.

BILLINGS, M. P., 1928a, The petrology of the North Conway Quadrangle in the White Mountains of New Hampshire: *Am. Acad. Arts and Sci. Proc.*, **63**, 67–138.

————, 1928b, The chemistry, optics, and genesis of the hastingsite group of amphiboles: *Am. Mineralogist.*, **13**, 287–296.

BINNS, R. A., 1965a, Hornblendes from some basic hornfelses in the New England region, New South Wales: *Mineralog. Mag.*, **34**, 52–65.

————, 1965b, The mineralogy of metamorphosed basic rocks from the Willyama Complex, Broken Hill district, New South Wales, Part I. Hornblendes: *Mineralog. Mag.*, **35**, 306–326.

BIRCH, F., and LeCOMTE, P., 1960, Temperature-pressure plane for albite composition: *Am. Jour. Sci.*, **258**, 209–217.

BLOXAM, T. W., 1956, Jadeite-bearing metagraywackes in California: *Am. Mineralogist*, **41**, 488–496.

————, 1959, Glaucophane-schists and associated rocks near Valley Ford, California: *Am. Jour. Sci.*, **257**, 95–112.

————, 1960, Jadeite-rocks and glaucophane-schists from Angel Island, San Francisco Bay, California: *Am. Jour. Sci.*, **258**, 555–573.

BORG, I. Y., 1956, Glaucophane schists and eclogites near Healdsburg, California: *Geol. Soc. America Bull.*, **67**, 1563–1584.

————, 1967, Optical properties and cell parameters in the glaucophane-riebeckite series: *Contr. to Min. Pet., Beit. Min. u. Pet.*, **15**, 67–92.

BORLEY, G. D., 1963, Amphiboles from the Younger Granites of Nigeria. Part I. Chemical Classification: *Mineralog. Mag.*, **33**, 358–376.

————, and Frost, M. T., 1963, Some observations on igneous ferrohastingsites: *Mineralog. Mag.*, **33**, 646–662.

BOWEN, N. L., 1940, Progressive metamorphism of siliceous limestone and dolomites: *Jour. Geology*, **48**, 225–274.

————, and SCHAIRER, J. F., 1935, Grunerite from Rockport, Massachusetts, and a series of synthetic fluor-amphiboles: *Am. Mineralogist*, **20**, 543–551.

————, and TUTTLE, O. F., 1949, The system MgO-SiO₂-H₂O: *Geol. Soc. America Bull.*, **60**, 439–460.

BOYD, F. R., 1959, Hydrothermal investigations of amphiboles: 377–396 in Researches in geochemistry: P. H. Abelson, ed.: John Wiley and Sons, Inc., New York, 511 pp.

————, and ENGLAND, J. L., 1960, Apparatus for phase-equilibrium measurements at pressures up to 50 kilobars and temperatures up to 1750° C: *Jour. Geophy. Res.*, **65**, 741–748.

————, and SCHAIRER, J. F., 1964, The system $MgSiO_3$-$CaMgSi_2O_6$: *Jour. Petrology*, **5**, 275–309.

BUDDINGTON, A. F., and LEONARD, B. F., 1953, Chemical petrology and mineralogy of hornblendes in northwest Adirondack granitic rocks: *Am. Mineralogist*, **38**, 891–902.

BUGGE, J. A. W., 1943, Geological and petrological investigations in the Kongsberg-Bamble formation: *Norges Geol. Undersok*, **160**, 150 p.

BURNS, R. G., 1966, Electronic spectra crystal-field phenomena, and iron-magnesium ratios in coexisting pyroxenes and amphiboles: (1966 London IMA-abstract).

————, and STRENS, R. G. J., 1966, Infrared study of the hydroxyl bands in clinoamphiboles: *Science*, **153**, 890–892.

CHAPMAN, R. W., and WILLIAMS, C. R., 1935, Evolution of the White Mountain magma series: *Am. Mineralogist*, **20**, 502–530.

CHOUDHURI, A., and WINKLER, H. G. F., 1967, Anthophyllit und Hornblende in einigen metamorphen Reaktionen: *Contr. to Min. Pet., Beit. Min. u. Pet.*, **14**, 293–315.

CLARK, S. P., 1957, A note on calcite-aragonite equilibrium: *Am. Mineralogist*, **42**, 564–566.

————, and RINGWOOD, A. E., 1964, Density distribution and constitution of the mantle: *Rev. Geophysics*, **2**, 35–88.

COLEMAN, R. G., and LEE, D. E., 1962, Metamorphic aragonite in the glaucophane schists of Cazadero, California: *Am. Jour. Sci.*, **260**, 577–593.

————, and LEE, D. E., 1963, Glaucophane-bearing metamorphic rock types of the Cazadero area, California: *Jour. Petrology*, **4**, 260–301.

COLLINS, R. S., 1942, Cummingtonite and gedrite from Sutherland: *Mineralog. Mag.*, **26**, 254–259.

COLVILLE, A. A., and GIBBS, G. V., 1965, Refinement of the crystal structure of riebeckite: *Geol. Soc. America*, Special Paper, **82**, 31 (abstract).

————, manuscript in preparation (β variation).

COLVILLE, P., ERNST, W. G., and GILBERT, M. C., 1966, Relationships between cell parameters and chemical compositions of monoclinic amphiboles: *Am. Mineralogist*, **51**, 1727–1754.

COMEFORO, J. E., and KOHN, J. A., 1954, Synthetic asbestos investigations, I: Study of synthetic fluor-tremolite: *Am. Mineralogist*, **39**, 537–548.

Compton, R. R., 1958, Significance of amphibole paragenesis in the Bidwell Bar region, California: *Am. Mineralogist*, **43**, 890–907.

Crawford, W. A., and Fyfe, W. S., 1965, Lawsonite equilibria: *Am. Jour. Sci.*, **263**, 262–270.

Cristophe-Michel-Lévy, 1957, Premiers stades du metamorphisme artificiel d'une dolomie siliceuse: formation de tremolite et de diopside: *Bull. Soc. franc. Min. Crist.*, **80**, 297–302.

Deer, W. A., Howie, R. A., and Zussman, J., 1963, Rock-forming minerals, **2**, Chain silicates: John Wiley and Sons, Inc., New York, 379 pp.

Dengo, G., 1953, Geology of the Caracas region, Venezuela: *Geol. Soc. America Bull.*, **64**, 7–40.

Doe, B. R., 1962, Relationships of lead isotopes among granites, pegmatites, and sulfide ores near Balmat, New York: *Jour. Geophys. Research*, **67**, 2895–2906.

Engel, A. E. J., and Engel, C. E., 1960, Progressive metamorphism and granitization of the major paragneiss, northwest Adirondack Mountains, New York: *Geol. Soc. America Bull.*, **71**, 1–58.

————, ————, 1962, Hornblendes formed during progressive metamorphism of amphibolites, northwest Adirondack Mountains, New York: *Geol. Soc. America Bull.*, **73**, 1499–1515.

Engel, C. E., 1959, Igneous rocks and constituent hornblendes of the Henry Mountains, Utah: *Geol. Soc. America Bull.*, **70**, 951–980.

Ernst, W. G., 1960, Stability relations of magnesioriebeckite: *Geochim. et Cosmochim. Acta*, **19**, 10–40.

————, 1961, Stability relations of glaucophane: *Am. Jour. Sci.*, **259**, 735–765.

————, 1962, Synthesis, stability relations, and occurrence of riebeckite and riebeckite-arfvedsonite solid solutions: *Jour. Geol.*, **70**, 689–736.

————, 1963a, Polymorphism in alkali amphiboles: *Am. Mineralogist*, **48**, 241–260.

————, 1963b, Petrogenesis of glaucophane schists: *Jour. Petrology*, **4**, 1–30.

————, 1965, Mineral parageneses in Franciscan metamorphic rocks, Panoche Pass, California: *Geol. Soc. America Bull.*, **76**, 879–914.

————, 1966, Synthesis and stability relations of ferrotremolite: *Am. Jour. Sci.*, **264**, 37–65.

————, and Seki, Y., 1967, Petrologic comparison of the Franciscan and Sanbagawa metamorphic terranes: *Jour. Tectonophys, in press*.

Eskola, P., 1914, On the petrology of the Orijarvi region in southwestern Finland: *Bull. Comm. geol. Finlande*, **40**, 279 pp.

————, 1929, Om Mineralfacies: *Geol. Foren. Stockh. Forh.*, **51**, 157–172.

————, 1950, Paragenesis of cummingtonite and hornblende from Muuruvesi, Finland: *Am. Mineralogist*, **35**, 729–734.

————, and Kervinen, T., 1936, A paragenesis of gedrite and cummingtonite from Isopaa in Kolvola, Finland: *Bull. Comm. geol. Finlande*, **115**, 475–487.

Eugster, H. P., 1957, Heterogeneous reactions involving oxidation and reduction at high pressures and temperatures: *Jour. Chem. Physics*, **26**, 1760–1761.

————, 1959, Reduction and oxidation in metamorphism: 397–426 in Researches in geochemistry: P. H. Abelson, ed.: John Wiley and Sons, Inc., New York, 511 pp.

————, and WONES, D. R., 1962, Stability relations of the ferruginous biotite, annite: *Jour. Petrology*, **3**, 82–125.

————, SKIPPEN, G. B., and HUEBNER, J. S., 1966, Experimental buffering systems for the control of gas fugacities in complex gas mixtures: *Trans. Am. Geophy. Union*, **47**, 211 (abstract).

————, ————, 1967, Igneous and metamorphic reactions involving gas equilibria: in Researches in geochemistry, vol. **2**: P. H. Abelson, ed.: John Wiley and Sons, Inc., New York, *in press*.

Evans, B. W., 1965, Application of a reaction-rate method to the breakdown equilibria of muscovite and muscovite plus quartz: *Am. Jour. Sci.*, **263**, 647–667.

FINGER, L. W., and ZOLTAI, T., 1967, Cation distribution in grunerite: *Trans. Am. Geophys. Union*, **48**, 233–234 (abstract).

FISCHER, K. F., 1966, A further refinement of the crystal structure of cummingtonite, $(Mg,Fe)_7(Si_4O_{11})_2$: *Am. Mineral.*, **51**, 814–818.

FLASCHEN, S. S., and OSBORN, E. F., 1957, Studies of the system iron oxide-silica-water at low oxygen partial pressures: *Econ. Geology*, **52**, 923–943.

FLOOR, P., 1966, Petrology of an aegirine-riebeckite gneiss-bearing part of the Hesperian Massif: the Galineiro and surrounding areas, Vigo, Spain: *Leidse Geol. Meded.*, **36**, 1–204.

FRANCIS, G. H., 1958, Petrological studies in the Glen Urquhart, Inverness-shire: *Bull. Brit. Mus. (Nat. Hist.), Min.*, v. **1**, no. **5**, 121–162.

FRENCH, B. M., 1966, Some geological implications of equilibrium between graphite and a C-H-O gas phase at high temperatures and pressures: *Rev. Geophys.*, **4**, 223–253.

————, and EUGSTER, H. P., 1965, Experimental control of oxygen fugacities by graphite-gas equilibriums: *Jour. Geophys. Res.*, **70**, 1529–1539.

FROST, M. T., 1963, Amphiboles from the Younger Granites of Nigeria. Part II. X-ray data: *Mineralog. Mag.*, **33**, 377–384.

FYFE, W. S., 1960, Hydrothermal synthesis and determination of equilibrium between minerals in the subsolidus region: *Jour. Geol.*, **68**, 553–566.

————, 1962, On the relative stability of talc, anthophyllite, and enstatite: *Am. Jour. Sci.*, **260**, 460–466.

————, TURNER, F. J., and VERHOOGEN, J., 1961, Coupled reactions in metamorphism: a correction: *Geol. Soc. America Bull.*, **72**, 169–170.

GHENT, E. D., 1965, Glaucophane-schist facies metamorphism in the Black Butte, area, Northern Coast Ranges, California: *Am. Jour. Sci.*, **263**, 385–400.

GHOSE, S., and HELLNER, E., 1959, The crystal structure of grunerite and observations on the Mg-Fe distribution: *Jour. Geol.*, **67**, 691–701.

————, 1961, The crystal structure of a cummingtonite: *Acta. Cryst.*, **14**, 622–627.

GIBBS, G. V., MILLER, J. L., and SHELL, H. R., 1962, Synthetic fluor-magnesio-richterite: *Amer. Mineralogist*, **47**, 75–82.

————, and PREWITT, C. T., 1966, Amphibole cation site disorder: (London IMA-abstract).

————, 1965, Crystal structure of protoamphibole: *Geol. Soc. America*, Special Paper, **82**, 71–72 (abstract).

GIER, T. E., COX, N. L., and YOUNG, H. S., 1964, The hydrothermal synthesis of sodium amphiboles: *Inorganic Chemistry*, **3**, 1001–1004.

GILBERT, M. C., 1966, Synthesis and stability relationships of ferropargasite: *Am. Jour. Sci.*, **264**, 698–742.

GREEN, D. H., and RINGWOOD, A. E., 1963, Mineral assemblages in a model mantle composition: *Jour. Geophys. Res.*, **68**, 937–945.

GREEN, J., 1959, Geochemical table of the elements for 1959: *Geo. Soc. America Bull.*, **70**, 1127–1184.

GREENWOOD, H. J., 1961, The system NaAlSi$_2$O$_6$-H$_2$O-Argon: total pressure and water pressure in metamorphism: *Jour. Geophys. Res.*, **66**, 3923–3946.

————, 1962a, Metamorphic reactions involving two volatile components: *Carnegie Inst. Wash. Yearbook*, **61**, 82–85.

————, 1962b, Synthesis and stability of anthophyllite: *Carnegie Inst. Wash. Yearbook*, **61**, 85–88.

————, 1963, The synthesis and stability of anthophyllite: *Jour. Petrology*, **4**, 317–351.

————, 1967, Mineral equilibria in the system MgO-SiO$_2$-H$_2$O-CO$_2$: in Researches in Geochemistry, **2**: P. H. Abelson, ed.: John Wiley and Sons, Inc., New York, *in press*.

GRIGGS, D. T., and KENNEDY, G. C., 1956, A simple apparatus for high pressures and temperatures: *Am. Jour. Sci.*, **254**, 722–735.

GUNDERSON, J. N., and SCHWARTZ, G. M., 1962, The geology of the metamorphosed Biwabik iron-formation, Eastern Mesabi District, Minnesota: *Minn. Geol. Survey Bull.*, **43**, 139 p.

HALL, A. L., 1930, Asbestos in the Union of South Africa: *Geol. Survey Union of South Africa*, Mem. **12**, 329 pp.

HALLIMOND, A. F., 1943, On the graphical representation of the calciferous amphiboles: *Am. Mineralogist*, **28**, 65–89.

HARRY, W. T., 1950, Aluminum replacing silicon in some silicate lattices: *Mineralog. Mag.*, **29**, 142–149.

HART, S. R., and ALDRICH, L. T., 1967, Fractionation of potassium/rubidium by amphiboles: implications regarding mantle composition: *Science*, **155**, 325–327.

HELLNER, E., HINRICHSEN, Th., and SEIFERT, F., 1965, The study of mixed crystals of minerals in metamorphic rocks: 155–168 in Controls of metamorphism, W. S. Pitcher and G. W. Flinn, eds.: John Wiley and Sons, Inc., New York, 368 pp.

————, and SCHÜRMANN, K., 1966, Stability of metamorphic amphiboles: the tremolite-ferroactinolite series: *Jour. Geol.*, **74**, 322–331.

HELLNER, E., and SCHÜRMANN, K., 1967, Stability of metamorphic amphiboles: the tremolite-ferrotremolite series: a reply: *Jour. Geology*, **75**, 351.

HERITSCH, H., PAULITSCH, P. and WALITZI, E. M., 1957, Die Struktur von Karinthin und einer barroisitschen Hornblende: *Tschermaks Mineral. Petrol. Mitt.*, **6**, 215–225.

————, BERTOLDI, G., and WALITZI, E. M., 1960, Strukturuntersuchung in einer basaltischen Hornblende vom Kuruzzenkogel, südlich Fehring, Steiermark: *Tschermaks Mineral. Petrog. Mitt.*, **7**, 210–217.

————, and KAHLER, E., 1960, Strukturuntersuchung an zwei Kluftkarinthinen, Ein Beitrag zur Karinthinfrage: *Tschermaks Mineral. Petrog. Mitt.*, **7**, 218–234.

————, and RIECHERT, L., 1960, Strukturuntersuchung an einer basaltischen Hornblende von Cernosin, CSR: *Tschermaks Mineral. Petrog. Mitt.*, **7**, 235–245.

HIDE, K., 1961, Geologic structure and metamorphism of the Sanbagawa crystalline schists of the Besshi-Shirataki Mining district in Shikoku, southwest Japan: *Geol. Rept. Hiroshima Univ.*, **9**, 1–87.

HIETANEN, A., 1959, Kyanite-garnet gedritite near Orofino, Idaho. *Am. Mineralogist*, **44**, 539–564.

HINRICHSEN, TH. J., 1966, Hydrothermal investigation and stability relations of orthorhombic amphiboles: (1966 London IMA-abstract).

HSU, L. C., *in press*, Selected phase relationships in the system Al-Mn-Fe-Si-O-H, a model for garnet equilibria: *Jour. Petrology*.

HUANG, W. T., 1958, Riebeckite granite in the Wichita Mountains, Oklahoma: *Geol. Soc. America Bull.*, **69**, 1191–1192.

HUTTON, C. O., 1940, Metamorphism in the Lake Wakatipu region western Otago, New Zealand: Dept. Sci. Ind. Res. N. S. Geol. Mem., **5**, 90 p.

IWASAKI, M., 1963, Metamorphic rocks of the Kotu-Bizan area, eastern Shikoku: *Jour. Fac. Sci., Univ. Tokyo, Sec. II*, **15**, 1–90.

JACOBSON, R. R. E., MACLEOD, W. N., and BLACK, R., 1958, Ring-complexes in the younger granite province of Northern Nigeria: *Geol. Soc. London Mem. 1*, 71 p.

KAWAHARA, A., 1963, X-ray studies on some alkaline amphiboles: *Mineralog. Jour.*, **4**, 30–40.

KLEIN, C., 1964, Cummingtonite-grunerite series: a chemical, optical and x-ray study: *Am. Mineralogist*, **49**, 963–982.

————, 1966, Mineralogy and petrology of the metamorphosed Wabush iron formation, Southwestern Labrador: *Jour. Petrology*, **7**, 246–305.

————, *in press*, Coexisting amphiboles: *Jour. Petrology*.

KOHN, J. A., and COMEFORO, J. E., 1955, Synthetic asbestos investigations II. X-ray and other data on synthetic fluor-richterite-edenite and boron edenite: *Am. Mineralogist*, **40**, 410–421.

KRANCK, S. H., 1961, A study of phase equilibria in a metamorphic iron formation: *Jour. Petrology* **2**, 137–184.

LARSEN, E. S., 1942, Alkalic rocks of Iron Hill, Gunnison County, Colorado: *U. S. Geol. Surv. Prof.* Paper no. **197A**, 1–64.

LEAKE, B. E., 1962, On the non-existence of a vacant area in the Hallimond calciferous amphibole diagram: *Japanese Jour. Geol. Geog.*, **33**, 1–13.

————, 1965, The relationship between composition of calciferous amphibole and grade of metamorphism: 299–318 in Controls of metamorphism, W. S. Pitcher and G. W. Flinn, eds.: John Wiley and Sons, Inc., New York, 368 pp.

References   **111**

LEE, D. E., COLEMAN, R. G., BASTRON, H., and SMITH, V. C., 1966, A two-amphibole glaucophane schist in the Franciscan Formation, Cazadero area, Sonoma County, California: *U. S. Geol. Survey Prof.* Paper 550-C, c148–c157.

LINDSLEY, D. H., 1967, The join hedenbergite-ferrosilite at high pressures and temperatures: *Carnegie Inst. Wash. Yearbook*, **65**, 230–232.

LIPMAN, P. W., 1964, Mineralogy and paragenesis of amphiboles from Gibson Peak pluton, Northern California: *Am. Mineralogist*, **49**, 1321–1330.

LUTH, W. C., and TUTTLE, O. F., 1963, Externally heated cold-seal pressure vessels for use to 10,000 bars and 750° C: *Am. Mineralogist*, **48**, 1401–1403.

MACKAY, R. A., GREENWOOD, R., and ROCKINGHAM, J. E., 1949, The geology of the Plateau tinfields, Re-survey 1945–48: *Geol. Survey Nigeria Bull.*, **19**.

MACKENZIE, W. S., 1965, Some comments on the application of experimental results to the study of metamorphism: 268–273 in Controls of Metamorphism, W. S. Pitcher and G. W. Flinn, eds.: John Wiley and Sons, Inc., New York, 368 pp.

MASON, B., 1966, Pyrope, augite and hornblende from Kakanui, New Zealand: *New Zealand Jour. Geol. and Geophys.*, **9**, 474–480.

MCKEE, B., 1962a, Widespread occurrence of jadeite, lawsonite, and glaucophane in central California: *Am. Jour. Sci.*, **260**, 596–610.

———, 1962b, Aragonite in the Franciscan rocks of the Pacheco Pass area, California: *Am. Mineralogist*, **47**, 379–387.

Melson, W. G., JAROSEWICH, E., BOWEN, V. T., and THOMPSON, G., 1967, St. Peter and St. Paul Rocks: a high-temperature, mantle-derived intrusion: *Science*, **155**, 1532–1535.

METZ, P. W., and WINKLER, H. G. F., 1964, Experimentelle Untersuchung der Diopsidbildung aus Tremolit, Calcit und Quarz: *Naturwissenschaften*, **51**, 460.

METZ, P., 1967, Die obere Stabilitätsgrenze von Tremolit bei der Metamorphose von kieseligen Karbonaten: *Contr. to Min. Pet., Beit. Min. u. Pet.*, **15**, 272–280.

MILES, K. R., 1942, The blue asbestos bearing banded iron formations of the Hamersley Ranges, Western Australia: *Geol. Survey Western Australia*, Bull. 100, pt. 1, 5–37.

MILTON, C., and EUGSTER, H. P., 1959, Mineral assemblages of the Green River Formation: 118–150 in Researches in geochemistry: P. H. Abelson, ed.: John Wiley and Sons, Inc., New York, 511 pp.

MIYASHIRO, A., 1957, The chemistry, optics, and genesis of the alkali-amphiboles: *Jour. Fac. Sci. Univ. Tokyo*, Sec. II, **11**, 57–83.

———, 1958, Regional metamorphism of the Gosaisyo-Takanuki district in the Central Abukuma Plateau: *Jour. Fac. Sci., Univ. Tokyo*, Sec. II, **11**, 219–272.

———, 1961, Evolution of metamorphic belts: *Jour. Petrology*, **2**, 277–311.

———, 1964, Oxidation and reduction in the Earth's crust with special reference to the role of graphite: *Geochim. et Cosmochim. Acta*, **28**, 717–729.

———, and BANNO, S., 1958, Nature of glaucophanitic metamorphism: *Am. Jour. Sci.*, **256**, 97–110.

MUELLER, R. F., 1960, Compositional characteristics and equilibrium relations in mineral assemblages of a metamorphosed iron formation: *Am. Jour. Sci.*, **258**, 449–497.

————, 1962, Energetics of certain silicate solid solutions: *Geochim. et. Cosmochim. Acta*, **26**, 581–598.

————, 1967, Stability of metamorphic amphiboles: the tremolite-ferroactinolite series: a discussion: *Jour. Geology*, **75**, 234–236.

NICHOLLS, G. D., and ZUSSMAN, J., 1955, The structural formula of a hydrous amphibole: *Mineralog. Mag.*, **30**, 717–722.

OLSEN, E., 1967, Amphibole: first occurrence in a meteorite: *Science*, **156**, 61–62.

ONUKI, H., 1964, Hornblendes from the ultramafic intrusives in the Kitakami Mountainland: *Jour. Japan Assoc. Min. Pet. Econ. Geol.*, **51**, 210–222.

ORVILLE, P. M., 1963, Alkali ion exchange between vapor and feldspar phases: *Am. Jour. Sci.*, **261**, 201–237.

OXBURGH, E. R., 1964, Petrological evidence of the presence of amphibole in the upper mantle and its petrogenetic and geophysical implications: *Geol. Mag.*, **101**, 1–19.

PALACHE, C., and WARREN, C. H., 1911, The chemical composition and crystallization of parisite and a new occurrence of it in the granite pegmatites at Quincy, Mass., U.S.A.: *Am. Jour. Sci.*, 4th ser., **31**, 533–558.

PAPIKE, J. J., and CLARK, J. R., *in press*, The crystal structure and cation distribution of glaucophane: *Am. Mineralogist*.

PEACOCK, M. A., 1928, The nature and origin of the amphibole asbestos of South Africa: *Am. Mineralogist*, **13**, 241–285.

PHILLIPS, R., and ROWBOTHAM, G., 1967, Studies on synthetic alkali amphiboles: *Mineralog. Mag.*, *in press*.

PIRANI, R., 1952, I minerali del gruppo dell'Ortler-IV. Antofillite e actinolite de Val di Bresimo. Solubilita di questa actinolite: *Atti (Rend.) Accad. Naz, Lincei, Cl. Sci. fis. mat. nat.*, ser. 8, v. 13, sem. 2, 83–88.

PLAS, L. VAN DER, 1959, Petrology of the northern Adula region, Switzerland (with particular reference to glaucophane-bearing rocks): *Leidse Geol. Meded*, **24**, 415–602.

PREWITT, C. T., 1964, Crystal structures of two synthetic amphiboles: *Geol. Soc. America*, Special Paper, **76**, 132–133 (abstract).

PRIDER, R. T., 1944, The geology and petrology of part of the Toodyay district, Western Australia: *Jour. Roy. Soc. W. Australia*, **28**, 83–137.

RABBITT, J. C., 1948, A new study of the anthophyllite series: *Am. Mineralogist*, **33**, 263–323.

RAMBERG, H., 1944, Petrological significance of subsolidus phase transitions in mixed crystals: *Norsk geol. tidsskr.*, **24**, 42–74.

RINGWOOD, A. E., 1966, Mineralogy of the Mantle: 357–399 in Advances in Earth Science: P. M. Hurley, ed.: The M.I.T. Press, Cambridge, Mass., 502 pp.

ROBERTSON, E. C., BIRCH, F., and MACDONALD, G. J. F., 1957, Experimental determination of jadeite stability relations to 25,000 bars: *Am. Jour. Sci.*, **255**, 115–137.

ROEVER, W. P. DE, 1947, Igneous and metamorphic rocks in eastern Celebes: 65–173 in Geological exploration in the Island of Celebes: North Holland Publishing Co., Amsterdam, 346 pp.

————, 1955, Genesis of jadeite by low-grade metamorphism: *Am. Jour. Sci.*, **253**, 283–298.

ROSENFELD, J. L., 1961, The contamination-reaction rules: *Am. Jour. Sci.*, **259**, 1–23.

SCHÜRMANN, K., 1966, Synthesis and stability field of cummingtonite: (London, IMA-abstract).

SCLAR, C. B., CARRISON, L. C. and STEWART, O. M., 1967, High-pressure synthesis of a new hydroxylated pyroxene in the system MgO-SiO_2-H_2O: *Trans. Am. Geophy. Union*, **48**, 226 (abstract).

SEITSAARI, J., 1952, On association of cummingtonite and hornblende: *Ann. Acad. Sci. Fenn.*, Ser. A, **III**, no. 30, 20 p.

SEKI, Y., 1958, Glaucophanitic regional metamorphism in the Kanto Mountains, Central Japan: *Japanese Jour. Geol. Geog.*, **29**, 233–258.

————, 1960, Jadeite in Sanbagawa crystalline schists of Central Japan: *Am. Jour. Sci.*, **258**, 705–715.

SHELL, H. R., COMEFORO, J. E., and EITEL, W., 1958, Synthetic asbestos investigations: synthesis of fluoramphiboles from melts: *U. S. Bur. Mines, Rept. Inves.* 5417.

SHIDO, F., 1958, Plutonic and metamorphic rocks of the Nakoso and Iritono districts in the Central Abukuma plateau: *Jour. Fac. Sci., Univ. Tokyo*, Sec. II, **11**, 131–217.

————, and MIYASHIRO, A., 1959, Hornblendes of basic metamorphic rocks: *Jour. Fac. Sci., Univ Tokyo*, Sec. II, **12**, 85–102.

SIMPSON, R. S., 1932, Contributions to the mineralogy of Western Australia—Series VII: *Jour. Roy. Soc. W. Australia*, **18**, 61–74.

SKIPPEN, G. B., 1967, An experimental study of the metamorphism of siliceous carbonate rocks: unpub. Ph.D. dissertation, The Johns Hopkins Univ., 251 pp.

SMITH, J. R., 1957, Reconnaissance in the system FeO-Fe_2O_3-SiO_2-H_2O; *Carnegie Inst. Wash. Yearbook*, **56**, 230–231.

SØRENSON, H., 1955, On sapphirine from west Greenland: *Medd. om Gronland*, **137**, No. 1, 3–32.

SUBRAMANIAM, A. P., 1956, Mineralogy and petrology of the Sittampundi complex, Salem District, Madras State, India: *Geol. Soc. America Bull.*, **67**, 317–390.

SUNDIUS, N., 1945, The position of the richterite in the amphibole group: *Geol. For. Forh.*, Stockholm, **67**, 266–270.

————, 1946, The classification of the hornblendes and the solid solution relations in the amphibole group: *Arsbok Sveriges Geol. Undersok.*, **40**, no. 4, 1–36.

SUZUKI, J., 1930, Petrological study of the crystalline schist system of Shikoku, Japan: *Jour. Fac. Sci., Hokkaido Imp. Univ.*, Ser. IV, **1**, 27–111.

SWITZER, G., 1951, Mineralogy of the California glaucophane schists: *Calif. Div. Mines Bull.*, **161**, 51–70.

THOMPSON, J. B., Jr., 1955, The thermodynamic basis for the mineral facies concept: *Am. Jour. Sci.*, **253**, 65–103.

TILLEY, C. E., 1937, The paragenesis of kyanite amphibolites: *Mineralog. Mag.*, **24**, 555–568.

————, 1938, The status of hornblende in low grade metamorphic zones of green schists: *Geol. Mag.*, **75**, 497–511.

————, 1957, Paragenesis of anthophyllite and hornblende from the Bancroft area, Ontario: *Am. Mineralogist*, **42**, 412–416.

TURNER, F. J., 1935, Contribution to the interpretation of mineral facies in metamorphic rocks: *Am. Jour. Sci.*, 4th ser., **29**, 409–421.

TURNOCK, A. C., and EUGSTER, H. P., 1962, The Fe-Al oxides: phase relationships below 1000° C: *Jour. Petrology*, **3**, 533–565.

TUTTLE, O. F., 1948, A new hydrothermal quenching apparatus: *Am. Jour. Sci.*, **246**, 628–635.

————, 1949, Two pressure vessels for silicate-water studies: *Geol. Soc. America, Bull.*, **60**, 1727–1729.

————, and BOWEN, N. L., 1958, Origin of granite in the light of experimental studies in the system $NaAlSi_3O_8$-$KAlSi_3O_8$-SiO_2-H_2O: *Geol. Soc. America*, Mem. 74, 153 pp.

UPTON, B. G. J., 1960, The alkaline igneous complex of Kungnat Fjeld, South Greenland: *Medd. om Gron.*, **123**, no. 4, 1–145.

VERNON, R. H., 1962, Coexisting cummingtonite and hornblende in an amphibolite from Duchess, Queensland, Australia: *Am. Mineralogist*, **47**, 360–370.

VILLIERS, J. E., DE, 1949, Note on an unusual amphibole from Zesfontein, Southwest Africa: *Trans. Geol. Soc. South Africa*, **51**, 77–80.

WARREN, B. E., 1929, The structure of tremolite: *Zeit. fur Krist.*, **72**, 42–57.

————, and MODELL, D. I., 1930, The structure of anthophyllite: *Zeit. Krist.*, **75**, 161–178.

WARREN, C. H., 1903, Mineralogical notes: *Am. Jour. Sci.*, 4th ser., **16**, 337–344.

WATTERS, W. A., 1959, An association of hornblende and cummingtonite from Ringaringa, Stewart Island, New Zealand: *New Zealand Jour. Geol. Geophys,*. **2**, 248–255.

WHITE, A. J. R., 1962, Aegirine-riebeckite schists from South Westland, New Zealand: *Jour. Petrology*, **3**, 38–48.

WHITTAKER, E. J. W., 1949, The structure of Bolivian crocidolite: *Acta Cryst.*, **2**, 312–317.

————, 1960, The crystal chemistry of the amphiboles: *Acta Cryst.*, **13**, 291–298.

————, 1966, Classification of the amphiboles: (1966 London IMA-abstract).

WINKLER, H. G. F., 1965, Petrogenesis of metamorphic rocks: Springer-Verlag New York Inc., N. Y., 220 pp.

WISEMAN, J. D. H., 1934, The central and southwest Highland epidiorites: a study in progressive metamorphism: *Quart. Jour. Geol. Soc. London*, **90**, 354–417.

WONES, D. R., and EUGSTER, H. P., 1965, Stability of biotite: experiment, theory, and application: *Am. Mineralogist*, **50**, 1228–1272.

WOODARD, H. H., 1957, Diffusion of chemical elements in some naturally occurring silicate inclusions: *Jour. Geol.*, **65**, 61–84.

WYMOND, A. P., and WILSON, R. L., 1951, An occurrence of crocidolite near Robertstown, South Australia: *Trans. Roy. Soc. South Australia*, **74**, 44–48.

YAGI, K., 1953, Petrochemical studies of the alkalic rocks of the Morotu district, Sakhalin: *Geol. Soc. America Bull.*, **64**, 769–909.

YAMAZAKI, T., ONUKI, H., and TIBA, T., 1966, Significance of hornblende gabbroic inclusions in calc-alkali rocks: *Jour. Japan. Assoc. Min. Pet. Econ. Geol.*, **55**, 87–103.

YODER, H. S., Jr., 1952, The MgO-Al₂O₃-SiO₂-H₂O system and the related metamorphic facies: *Am. Jour. Sci.*, Bowen Vol., 569–627.

————, and TILLEY, C. E., 1962, Origin of basalt magmas: an experimental study of natural and synthetic rock systems: *Jour. Petrology*, **3**, 342–532.

ZEN, E-AN, 1961, The zeolite facies: an interpretation: *Am. Jour. Sci.*, **259**, 401–409.

ZUSSMAN, J., 1955, The crystal structure of an actinolite: *Acta Cryst.*, **8**, 301–308.

————, 1959, A re-examination of the structure of tremolite: *Acta Cryst.*, **12**, 309–312.

AUTHOR INDEX

Ahlfeld, F., 97
Ahrens, L. H., 12
Akella, J., 44, 45, 48
Aldrich, L. T., 75
Appleman, D. E., 2

Bancroft, G. M., 31
Banno, S., 68, 69, 100
Barnes, H. L., 39
Barnes, V. E., 8
Bastron, H., (see Lee et al., 28)
Bertoldi, G. (see Heritsch, et al., 4)
Billings, M. P., 71
Binns, R. A., 69, 71
Birch, F., 97
Black, R. (see Jacobson et al., 100)
Bloxam, T. W., 97
Borg, I. Y., 12, 68
Borley, G. D., 71, 101
Bowen, N. L., 2, 39, 40, 45, 68, 74, 80, 101
Bowen, V. T. (see Melson et al., 75)
Boyd, F. R , 2, 13, 20, 34, 50, 51, 58, 60, 61, 66, 68, 73, 94, 102
Brown, G. M., 3
Buddington, A. F., 71
Bugge, J. A. W., 28
Burns, R. G., 31, 76

Carrison, L. C. (see Sclar et al., 76)
Chapman, R. W., 100
Choudhuri, A., 43
Christophe-Michel-Levy, 95
Clark, J. R., 4, 9, 95
Clark, S. P., 75–97
Coleman, R. G., 68, 97, 98
Collins, R. S., 28
Colville, A. A., 4, 7, 9, 15
Colville, P., 5, 10, 12, 13, 16, 19, 22, 24, 25, 27, 67, 95

Comeforo, J. E., 13, 67
Compton, R. R., 28, 71
Cox, N. L. (see Gier et al., 96)
Crawford, W. A., 97

Deer, W. A., 46
Dengo, G., 28
Doe, B. R., 48

Eitel, W. (see Shell et al., 67)
Engel, A. E. J., 48, 69
Engel, C. E., 48, 69, 71
England, J. L., 94
Ernst, W. G., 13, 14, 28, 31, 35, 54–58, 60, 68, 71, 75, 77–100, 102
Eskola, P., 28, 47, 48, 97
Eugster, H. P., 35, 39, 53, 54, 65, 97
Evans, B. W., 79

Finger, L. W., 4, 9
Fischer, K. F., 4, 9, 31
Flaschen, S. S., 45
Floor, P., 99
Francis, G. H., 68
French, B. M., 39
Frost, M. T., 71
Fyfe, W. S., 37, 39, 40, 97

Ghent, E. D., 97, 98
Ghose, S., 4, 9
Gibbs, G. V., 3, 4, 6, 7, 9, 14, 15, 76, 96
Gier, T. E., 96
Gilbert, M. C., 3, 13, 61–66, 71, 73, 102
Green, D. H., 74
Green, J., 12
Greenwood, H. J., 3, 14, 39–43, 48–50, 52–54, 60, 75, 80, 102

SUBJECT INDEX

MINERALS, ROCKS AND INORGANIC MATERIALS—*Series*

EXPERIMENTAL MINERALOGY—*Sub-Series*

Titles in Preparation:

BELL, P. M., S. W. RICHARDSON, and M. CH. GILBERT, Synthesis and Stability of the Aluminum Silicate Minerals.

BROWN, G. M. and D. H. LINDSLEY, Synthesis and Stability of Pyroxenes.

DOE, B. R., Lead Isotopes in Igneous Petrology [Sub-Series on "Isotopes in Geology"]

EUGSTER, H. P. and D. R. WONES, Synthesis and Stability of Micas.

FAWCETT, J. J., Synthesis and Stability of Chlorite and Serpentine.

LINDSLEY, D. H., Synthesis and Stability of Iron-Titanium Oxides.

POWELL, J. L. and G. FAURE, Strontium Isotopes.

HANSEN, E., Strain Facies.
(This volume is in the Series "Minerals, Rocks and Inorganic Materials" only. It is not part of the Sub-Series.)

The titles are subject to change.